Workshop
Numeracy

Workshop Numeracy

Harry Ogden

MPhil, CEng, FIProdE, MIMechE

Pitman

PITMAN PUBLISHING LIMITED
128 Long Acre London WC2E 9AN

Associated Companies
Pitman Publishing New Zealand Ltd, Wellington
Pitman Publishing Pty Ltd, Melbourne

© H Ogden 1984

First published in Great Britain 1984

ISBN 0 273 01898 1

Printed in Great Britain at The Pitman Press, Bath

Preface

This book provides a basic course of numeracy to meet the mathematics requirements of young people entering the first year of training in mechanical and electrical engineering, and in other associated vocational training schemes. It covers the subject requirements of the City and Guilds of London Institute craft courses such as Basic Engineering Trade Subjects for students in courses outside Britain, in addition to providing for Basic Engineering Craft Studies courses for British students.

The subject material constitutes a basic unit of numeracy suitable for Vocational Preparation and first-year training courses for school leavers who have limited or no qualification at C.S.E. level. The examples and practical applications are also of value to students engaged on the first year of T.E.C. and other technician courses.

The treatment of the subject material is specifically designed for vocational students and features a comprehensive range of worked examples leading to graded exercises at the end of each chapter. Throughout the book a strong link between theory and industrial practice is developed through the use of workshop applications.

Finally, the author hopes that the book may be useful to mature craftsmen who feel a need to revise their basic technical knowledge or who are entering a re-training programme for additional skills.

I acknowledge with gratitude the unfailing support and unstinting help of my wife, Lily. My thanks go also to John Cushion of Pitman Books and George Tomlinson, for their professional advice and assistance, but most of all for their infectious enthusiasm which made working with them such a pleasure.

I dedicate this book on the behalf of many thousands of former engineering craft students to the memory of Ron Wood, a craft lecturer whose devotion to the interests of his students was an example to all his colleagues.

H. Ogden 1983

Contents

S.I. Units

S.I. is the abbreviation used in all languages for the International System of Units (Système International des Unités). The system has six arbitrary basic units.

Basic Units

QUANTITY	NAME OF UNIT	UNIT/SYMBOL
Time	second	s
Length	metre	m
Mass	kilogramme	kg
Temperature	kelvin	K
Electric current	ampere	A
Luminous intensity	candela	cd

Selected Derived Units

QUANTITY	NAME OF UNIT	UNIT/SYMBOL
Force	newton	N
Energy, work	joule	J
Power	watt	W
Temperature	degree Celsius	°C
Electric charge	coulomb	C
Electric potential	volt	V
Electric resistance	ohm	Ω
Frequency	hertz	Hz
Torque	newton metre	Nm
Velocity	metre per second	m/s
Area	square metre	m^2
Volume	cubic metre	m^3

Common Prefixes for S.I. Units

PREFIX	SYMBOL	MULTIPLYING FACTOR
mega	M	1 000 000
kilo	k	1 000
centi	c	0.01
milli	m	0.001
micro	μ	0.000 001

Useful Conversion Factors

MULTIPLY	BY	TO GIVE
inches	25.4	millimetres
feet	0.3048	metres
yards	0.9144	metres
miles	1.61	kilometres
square inches	645.2	square millimetres
square feet	0.0929	square metres
square yards	0.8361	square metres
cubic inches	16 390	cubic millimetres
cubic feet	0.0283	cubic metres
cubic yards	0.765	cubic metres
cubic metres	1 000	litres
gallons (Imperial)	4.546	litres
pounds (mass)	0.454	kilogrammes
feet per minute	0.0051	metres per second
horsepower	0.7457	kilowatts

1 Operations with Decimals

1.1 Accuracy in Workshop Calculations

Every craftsman must be competent at performing the calculations that arise in the normal operation of his trade. Engineering is a precision industry and, in many calculations relating to engineering production, accuracy is essential to ensure the successful assembly and functioning of the manufactured parts.

Many calculations that commonly occur in the workshop require only the application of the familiar four rules of number, i.e. addition, subtraction, multiplication, and division. The vital difference between real problems and textbook exercises is that the possible consequences of any error in workshop calculations could be the scrapping of expensive workpieces and the waste of many hours of skilled labour.

When attempting the problems at the end of this chapter, you should be aware that only the correct answer is acceptable, as the whole purpose of the exercise is to encourage a sense of responsibility for the result and to develop complete confidence in an ability to calculate with accuracy.

1.2 Addition of Decimals

When adding decimal quantities, place the numbers in columns so that the decimal points occur directly underneath one another and so that all figures having the same place value appear in the same column.

Example 1.1   Find the value of

3.518 + 1.64 + 5.047.

```
  3.518
  1.64
  5.047
  ─────
 10.205   (Ans.)
  ─────
```

Example 1.2   Calculate the overall length *L* of the turned shaft shown in fig. 1.1.

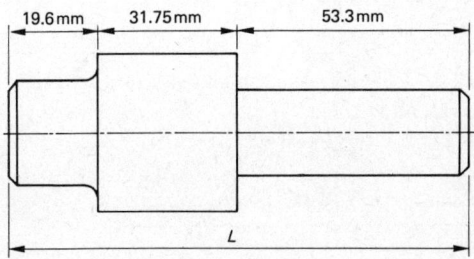

Fig. 1.1

Overall length of the shaft *L* is given by (19.6 + 31.75 + 53.3) mm.

```
  19.6
  31.75
  53.3
  ──────
 104.65      L = 104.65 mm   (Ans.)
  ──────
```

Example 1.3   The power used in kilowatts by four machine tools in a workshop is shown below.

centre lathe	1.46 kW
milling machine	2.35 kW
drilling machine	1.07 kW
power saw	0.94 kW

Find the total power consumption.

Total power consumed is

(1.46 + 2.35 + 1.07 + 0.94) kW.

```
  1.46
  2.35
  1.07
  0.94
  ────
  5.82
  ────
```

Total power consumed = 5.82 kW   (Ans.)

1.3 Subtraction of Decimals

When subtracting decimal quantities, use the same column arrangement as for addition, taking care to write the decimal points directly underneath one another.

Example 1.4 Find the value of $14.301 - 8.576$.

$$
\begin{array}{r}
14.301 \\
-\ 8.576 \\
\hline
5.725 \quad (Ans.) \\
\hline
\end{array}
$$

Example 1.5 Calculate the dimension l on the drilled plate shown in fig. 1.2.

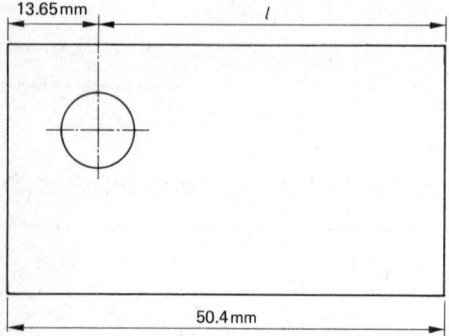

Fig. 1.2

The dimension l is given by $(50.4 - 13.65)$ mm.

$$
\begin{array}{r}
50.4 \\
-13.65 \\
\hline
36.75 \quad l = 36.75\,\text{mm} \quad (Ans.) \\
\hline
\end{array}
$$

Example 1.6 The power consumption of a machine tool motor is measured as 3.4 kW when rough machining and 1.85 kW when finish machining. Find the difference in power consumption of the two operations.

The difference in power consumption is given by $(3.4 - 1.85)$ kW.

$$
\begin{array}{r}
3.4 \\
-1.85 \\
\hline
1.55 \\
\hline
\end{array}
$$

Difference $= 1.55$ kW (*Ans.*)

1.4 Multiplication of Decimals

The number of decimal places in a decimal quantity is found by counting the number of figures to the right of the decimal point. When multiplying two decimal quantities, the number of decimal places in the answer will be the same as the total number of decimal places in both quantities.

The following method may be used when multiplying two decimal numbers:

Step (a) Disregard the decimal points and carry out the multiplication treating both numbers as whole numbers.

Step (b) Count the number of figures to the right of the decimal point in both numbers to give the total number of decimal places.

Step (c) Insert the decimal point in the answer so that the number of figures to the right of the point is the same as the total number of decimal places obtained in *Step (b)*.

Example 1.7 Find the value of 2.78×1.3.

Step (a): disregard the decimal points.

$$
\begin{array}{r}
278 \\
\times\ 13 \\
\hline
834 \\
2780 \\
\hline
3614 \\
\hline
\end{array}
$$

Step (b): the total number of decimal places in both numbers is $2 + 1 = 3$.

Step (c): insert the decimal point so that the answer has 3 decimal places:

3.614 (*Ans.*)

Example 1.8 Multiply 5.602 by 2.31.

$$
\begin{array}{r}
5602 \\
\times\ 231 \\
\hline
5602 \\
168060 \\
1120400 \\
\hline
1294062 \\
\hline
\end{array}
$$

Since the total number of decimal places $= 5$, then
12.940 62 (*Ans.*)

Example 1.9   Find the value of $1.3 \times 2.4 \times 3.7$.

$$
\begin{array}{r}
13 \\
\times\ 24 \\
\hline
52 \\
260 \\
\hline
312 \\
\times\ 37 \\
\hline
2184 \\
9360 \\
\hline
11544 \\
\hline
\end{array}
$$

Since the total number of decimal places = 3, then
11.544   (*Ans.*)

Example 1.10   Find the total height of a pile of five spacing washers each 3.65 mm thick.
Total height = 3.65×5 mm

$$
\begin{array}{r}
365 \\
\times\ 5 \\
\hline
1825 \\
\hline
\end{array}
$$

Since the total number of decimal places = 2, then
18.25 mm   (*Ans.*)

Example 1.11   Calculate the area of the rectangular plate shown in fig. 1.3.

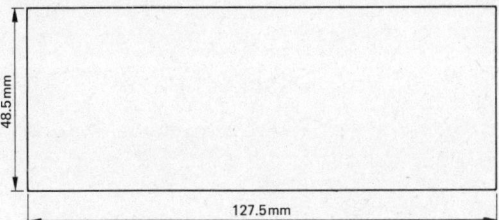

Fig. 1.3

Area of a rectangle = length × width
Area of the plate   = 127.5 mm × 48.5 mm

$$
\begin{array}{r}
1275 \\
\times\ 485 \\
\hline
6375 \\
10200 \\
510000 \\
\hline
618375 \\
\hline
\end{array}
$$

Since number of decimal places = 2, then
Area = 6 183.75 mm^2 (*Ans.*)

Example 1.12   Calculate the volume of a rectangular block of metal of length 9.5 cm, width 4.8 cm, and height 3.4 cm.

Volume of block = length × width × height
$$= 9.5\,\text{cm} \times 4.8\,\text{cm} \times 3.4\,\text{cm}$$
$$95 \times 48 \times 34 = 155040$$

Since total number of decimal places = 3, then
Volume = 155.040 cm^3 (*Ans.*)

Example 1.13   The voltage across a resistor in an electrical circuit can be found from the expression

Voltage (volts) = current (amperes) ×
resistance (ohms)

Find the voltage across a resistor of 3.5 ohms when the current flowing through the resistor is 0.7 amperes.

Voltage (volts) = current (amperes) ×
resistance (ohms)
$$= 0.7 \times 3.5$$
$$7 \times 35 = 245$$

Since total number of decimal places is 2, then
Voltage = 2.45 volts   (*Ans.*)

1.5 Division of Decimals

To make division by a decimal quantity easier, the divisor may be converted to a whole number by multiplying by 10 or by some multiple of 10. The dividend (the number to be divided) must also be multiplied by the same value so that the *ratio* between the divisor and the dividend remains constant. The sum may then proceed as an ordinary division sum without any adjustment to the answer.

When dividing by a decimal number the following method may be used:

Step (a)   Convert the divisor to a whole number by multiplying by 10, 100, 1 000, etc.
Step (b)   Multiply the dividend by the same value.
Step (c)   Proceed as an ordinary division sum.

Example 1.14   Find the value of 33.5 divided by 1.34, i.e.

$$\frac{33.5}{1.34}$$

Note   1.34 is the *divisor*
33.5 is the *dividend*

Step (*a*): convert the divisor to a whole number
  1.34 × 100 = 134
Step (*b*): multiply the dividend by the same value
  33.5 × 100 = 3 350
Step (*c*): divide 3 350 by 134

$$
\begin{array}{r}
25 \\
134\overline{)3350} \\
268 \\
\hline
670 \\
670 \\
\hline
\end{array}
$$

. . .

Therefore   33.5/1.34 = 25   (*Ans.*)

Example 1.15   Divide 74.39 by 4.3.

Step (*a*)   4.3 × 10 = 43
Step (*b*)   74.39 × 10 = 743.9
Step (*c*)   743.9 ÷ 43

$$
\begin{array}{r}
17.3 \\
43\overline{)743.9} \\
43 \\
\hline
313 \\
301 \\
\hline
129 \\
129 \\
\hline
\end{array}
$$

. . .

Therefore   743.9/43 = 17.3   (*Ans.*)

Example 1.16   How many pieces of length 83.5 mm can be sheared off a bar 1 419.5 mm long?

Number of pieces
= length of the bar ÷ length of a piece
= 1 419.5 ÷ 83.5
= 14 195 ÷ 835

$$
\begin{array}{r}
17 \\
835\overline{)14195} \\
835 \\
\hline
5845 \\
5845 \\
\hline
\end{array}
$$

. . . .   17 pieces   (*Ans.*)

Example 1.17   In one revolution a nut advances a distance of 3.18 mm along the length of a thread. How many revolutions must the nut make to advance a distance of 73.14 mm?

Number of revolutions required
= total distance ÷ distance per revolution
= 73.14 ÷ 3.18
= 7314 ÷ 318

$$
\begin{array}{r}
23 \\
318\overline{)7314} \\
636 \\
\hline
954 \\
954 \\
\hline
\end{array}
$$

. . .   23 revolutions   (*Ans.*)

Example 1.18   The face of a rectangular punch has an area of 825.5 mm². If the length of the face is 32.5 mm, calculate its width?

Area of a rectangle = length × width
        Width = area ÷ length
            = 825.5 ÷ 32.5
            = 8 255 ÷ 325

$$
\begin{array}{r}
25.4 \\
325\overline{)8255.0} \\
650 \\
\hline
1755 \\
1625 \\
\hline
1300 \\
1300 \\
\hline
\end{array}
$$

. . . .   Width = 25.4 mm   (*Ans.*)

Example 1.19   Given that 1 inch = 25.4 mm, convert the dimension 431.8 mm to inches.

Dimension in inches = 431.8 ÷ 25.4
          = 4 318 ÷ 254
          = 17

431.8 mm = 17 inches   (*Ans.*)

Example 1.20   The current flowing in an electrical circuit can be found from the expression

$$\text{Current (amperes)} = \frac{\text{potential difference (volts)}}{\text{resistance (ohms)}}$$

Calculate the current flowing in a conductor having a resistance of 2.5 ohms and connected to a 10.5 volt supply.

$$\text{Current (amperes)} = \frac{\text{potential difference (volts)}}{\text{resistance (ohms)}}$$

$$= \frac{10.5}{2.5} = \frac{105}{25} = 4.2$$

Current = 4.2 amperes   (*Ans.*)

1.6 Degree of Accuracy

In many workshop applications, it is not convenient or practical to work with values having a large number of decimal places.

For example: it is required to produce a replacement feedshaft for a lathe which was originally manufactured in Imperial or "inch" units. The diameter of the feedshaft is $1\frac{3}{8}$ inches.

To manufacture the shaft in metric units the diameter is converted to millimetres:

$$1\frac{3}{8} \times 25.4 = 34.925 \text{ mm}$$

The metric micrometer, which would be used to measure the finished diameter of the feedshaft, will only measure directly to an accuracy of 0.01 mm (approx. four ten-thousandths of an inch). Therefore, to dimension the diameter of the feedshaft as 34.925 mm is requiring the machinist to produce a workpiece which has a higher degree of accuracy than is normally obtainable with the standard measuring tools. Of course, such accuracy is completely unnecessary in this application. Thus for practical purposes, the metric dimension would be "rounded-off" to two decimal places to give 34.93 mm.

It is considered good practice, when making workshop calculations, to work out all values to one decimal place more than is required in the answer, and then to apply the procedure (shown in Section 1.7) to round-off the final answer to the required number of places.

The **degree of accuracy** required in a workshop calculation will normally be decided by practical considerations such as the limiting accuracy of the manufacturing processes involved. Answers should be shown to the number of decimal places consistent with the attainable degree of accuracy of the process or that required for the successful functioning of the manufactured component.

Answers such as approx. 45.7642 mm should be avoided. This dimension is much too precise to be called an approximation anyway and its degree of accuracy is far beyond what is attainable in any engineering process.

1.7 Decimal Places and Rounding-off

The following procedure is commonly used to **round-off** values in order to reduce the number of decimal places:

1 If the first figure after the required number of places is 5 or greater, then add 1 to the previous figure and omit the first figure.

2 If the first figure after the required number of places is less than 5, then simply omit the first figure without any change.

Thus, 3.087 correct to 2 decimal places is 3.09
3.084 correct to 2 decimal places is 3.08

Example 1.21 Show the value 2.4763 correct to *a*) 3 decimal places, *b*) 2 decimal places, *c*) 1 decimal place.

a) 2.4763 is 2.476 correct to 3 decimal places.
b) 2.4763 is 2.48 correct to 2 decimal places.
c) 2.4763 is 2.5 correct to 1 decimal place.

Example 1.22 A parting-off operation on a lathe can operate to a degree of accuracy of 0.1 mm. Rewrite the following dimensions in a form suitable for the operation: *a*) length 24.68 mm, *b*) length 32.563 mm, *c*) length 29.044 mm.

To 1 decimal place

a) 24.68 mm is written as 24.7 mm   (*Ans.*)
b) 32.563 mm is written as 32.6 mm   (*Ans.*)
c) 29.044 mm is written as 29.0 mm   (*Ans.*)

1.8 Significant Figures

A common workshop method of indicating the degree of accuracy of a value is to give the value correct to a number of **significant figures**.

For example, 32.57 by rounding-off and discarding the last figure becomes 32.6 correct to 3 significant figures.

Example 1.23 Round-off the following values to 3 significant figures: *a*) 13.76, *b*) 1.0104, *c*) 0.07658, *d*) 0.04863.

a) 13.76 is 13.8 to 3 s.f.   (*Ans.*)
b) 1.0104 is 1.01 to 3 s.f.   (*Ans.*)
c) 0.07658 is 0.0766 to 3 s.f.   (*Ans.*)
d) 0.04863 is 0.0486 to 3 s.f.   (*Ans.*)

Example 1.24 The table below shows the number of rivets produced by a machine each hour over a period of 6 hours continuous operation:

1	2	3	4	5	6
4951	5043	5039	4998	5022	4976

Rewrite the table giving the values correct to 2 significant figures.

1	2	3	4	5	6
5000	5000	5000	5000	5000	5000 (*Ans.*)

The rewritten table shows the rate of production to

the nearest 100 rivets, and what appeared as an irregular performance in the original table is now seen to be a steady production rate of about 5 000 rivets per hour.

Exercises 1

1.1 Find the value of
a) 41.2 + 8.95 + 17.06
b) 7.105 + 2.84 + 5.377
c) 4.9864 + 0.319 + 12.7205
d) 33.4 + 17.9 + 2.5 + 103.65
e) 2.0705 + 0.019 + 1.3508 + 0.0064

1.2 Find the value of   a) 89.64 − 37.95
b) 308.45 − 161.09   c) 17.05 − 9.318
d) 43.692 − 18.205   e) 172.408 − 93.0745

1.3 Find the value of   a) 3.19 × 1.4
b) 8.305 × 4.64   c) 12.45 × 17
d) 1.5 × 2.6 × 0.9   e) 135.5 × 59.4

1.4 Find the value of   a) 62.3 ÷ 0.7
b) 5.58 ÷ 1.8   c) 71.05 ÷ 2.9
d) 592.47 ÷ 9   e) 154.686 ÷ 25.4

1.5   Multiply 4.71 by 2.35 and give the answer correct to 2 decimal places.

1.6   Divide 59.0289 by 7 and give the answer correct to
a) 3 decimal places   b) 3 significant figures.

1.7   Round-off the following values:
a) 1.8927 to 3 decimal places
b) 0.041 to 2 decimal places
c) 8.065 16 to 4 decimal places
d) 79.676 to 4 significant figures
e) 0.035 37 to 3 significant figures

1.8   Calculate the overall length L of the shaft shown in fig. 1.4.

1.9   Calculate the overall length of the shaft shown in fig. 1.5.

1.10   Calculate the overall length of the pin shown in fig. 1.6.

1.11   Calculate the overall width of the gear shown in fig. 1.7.

1.12   Calculate the external dimensions A and B of the plate gauge shown in fig. 1.8.

1.13   Calculate the dimension l on the shaft shown in fig. 1.9.

1.14   Calculate the dimension C between the hole centres on the drilled plate shown in fig. 1.10.

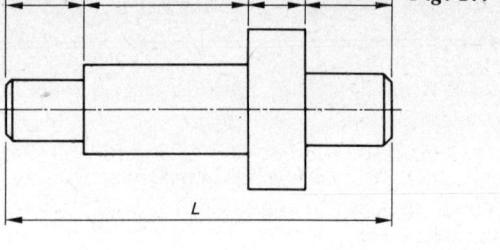
Fig. 1.4

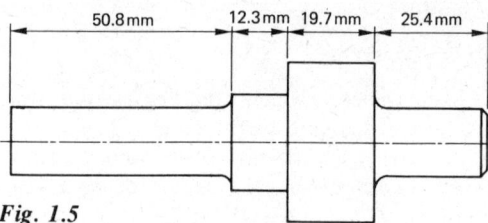

Fig. 1.5

Fig. 1.6

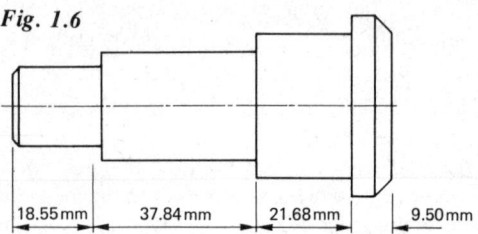

Fig. 1.7

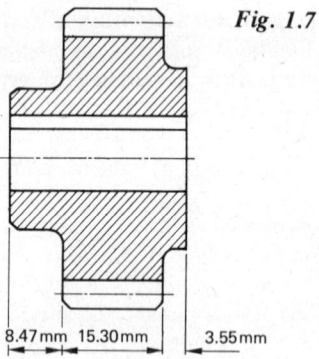

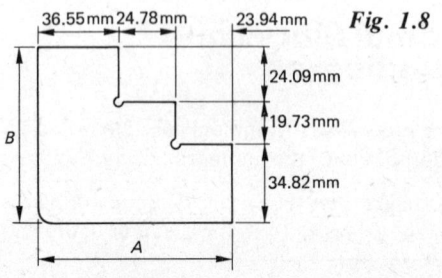
Fig. 1.8

6 Workshop Numeracy

1.15 Calculate the width of the slot in the fork end shown in fig. 1.11.

1.16 Calculate the dimensions a, b, c and d on the drilled plate shown in fig. 1.12.

1.17 The diameter of a circular bar is reduced in one cut in a lathe operation. The original diameter of the bar is 92.8 mm and the depth of cut is 1.24 mm. What is the finished diameter of the bar?

1.18 On the drilled strip shown in fig. 1.13 the pitch of the holes $P = 18.45$ mm. Calculate the dimension d.

1.19 Given that 1 inch = 25.4 mm, convert the following dimensions to millimetres:
a) 3 inches   *b*) 1.6 inches   *c*) 12.8 inches

1.20 Calculate the volume of the rectangular block of metal shown in fig. 1.14.
   (Volume = length × width × height.)

1.21 Calculate the area of a rectangle of length 64.4 mm and width 40.7 mm. Give the answer correct to 1 decimal place.
   (Area = length × width)

1.22 *a*) Calculate the total thickness of eight spacing washers each of thickness 5.35 mm.
b) How many of these washers would be required to make up a distance of 58.85 mm?

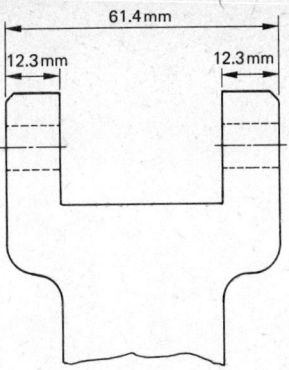

Fig. 1.11

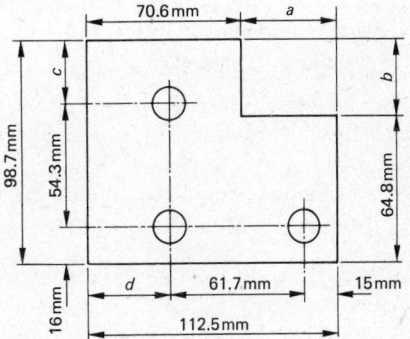

Fig. 1.12

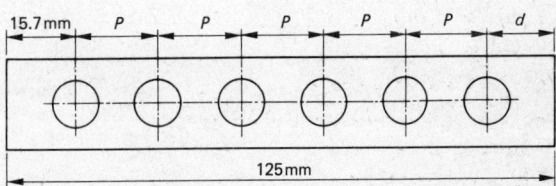

Fig. 1.13

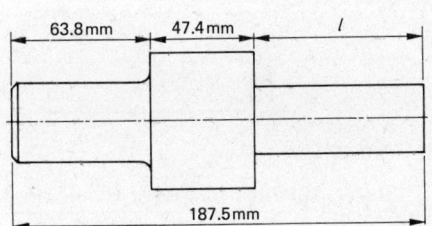

Fig. 1.9

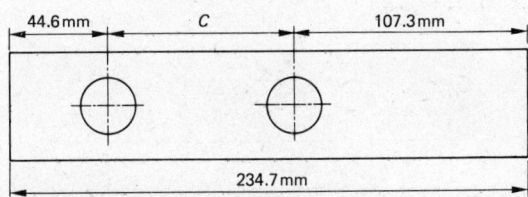

Fig. 1.10

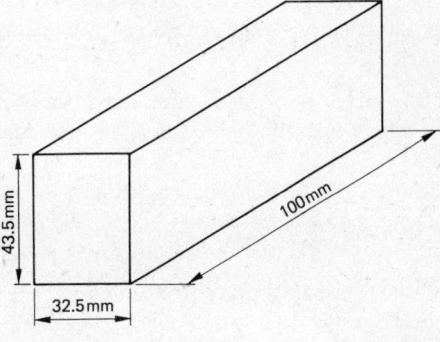

Fig. 1.14

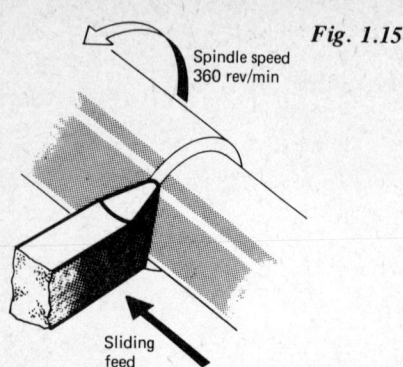

Fig. 1.15

Spindle speed
360 rev/min

Sliding
feed

1.23 During a sliding operation on a centre lathe (fig. 1.15) the cutting tool moves a distance of 108 mm along the work in 1 minute. If the spindle speed is 360 rev/min calculate the distance moved by the tool in one revolution of the spindle (i.e. feed/rev).

1.24 *a*) Nine revolutions advance a nut by a distance of 31.5 mm along the length of a thread. Calculate the pitch of the thread.
b) How many revolutions would be required to advance the nut through a distance of 66.5 mm?

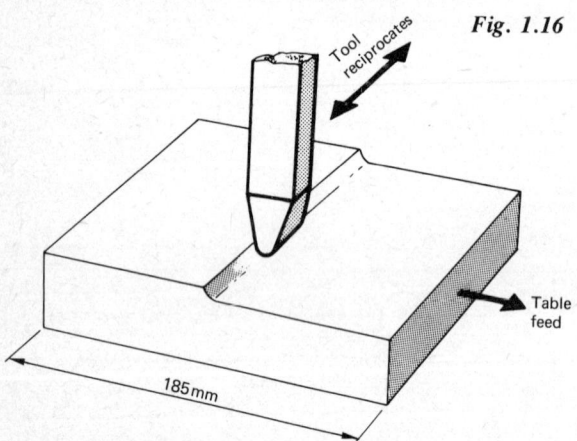

Fig. 1.16

Tool reciprocates

Table feed

185mm

1.25 Fig. 1.16 shows a block of metal of width 185 mm being machined on its upper surface in a shaping operation. If the table feed is 2.5 mm per stroke, calculate the number of strokes required to machine the surface.

1.26 Find the total power consumption in kilowatts of a workshop when operating the following machine tools:
Shaping machine 2.32 kW   Grinding machine 1.84 kW
Planing machine 4.75 kW   Band saw   0.92 kW
Bench drill   0.69 kW   Centre lathe   1.45 kW

1.27 What will be the total power consumption of the workshop in Exercise 1.26 when all the machines except the planing machine and the band saw are in use?

1.28 When resistors in an electrical circuit are connected in series, their equivalent resistance is given by the sum of their resistances. Thus the equivalent resistance R (ohms) of a number of resistors R_1, R_2, R_3, etc. is given by
$$R(\text{ohms}) = R_1 + R_2 + R_3 + \text{etc.}$$
a) Find the equivalent resistance of three resistors of 4.5, 7.5 and 15 ohms when connected in series.
b) An electrical circuit has four resistors connected in series. Three of the resistors have values of 8, 12.5 and 30 ohms. If the equivalent resistance of the circuit is 56 ohms, what is the value of the fourth resistor?

1.29 Given that
voltage (volts) = current (amperes)
$\times$ resistance (ohms)
find the voltage across
a) a resistor of 400 ohms when the current flowing is 0.35 amperes
b) a resistor of 7.5 ohms when the current flowing is 0.9 amperes
c) a resistor of 12.5 ohms when the current flowing is 1.5 amperes

1.30 Use the relationship
$$\text{Current (amperes)} = \frac{\text{potential difference (volts)}}{\text{resistance (ohms)}}$$
to complete the following table:

VOLTS	AMPERES	OHMS
110		12.5
210		40
60		8
12	1.2	
24	0.5	

1.31 Use the relationship
$$\text{Current (amperes)} = \frac{\text{quantity of charge (coulombs)}}{\text{time (seconds)}}$$
to complete the following table:

AMPERES	COULOMBS	SECONDS
	200	20
	10	0.2
3		10.5
40	180	
2.8		7.5

1.32 The storage capacity of a battery is given by
Capacity (ampere hours) = current (amperes) ×
time (hours)

Use this relationship to complete the table.

AMPERE HOURS	AMPERES	HOURS
	15	2.2
13	0.3	
37.5		2.75
27	2.2	
40	3.6	

1.33 Given that 1 kW = 1 000 W
a) Convert to kilowatts   i) 200 W
ii) 750 W   iii) 1 350 W
iv) 165 W   v) 75 W
b) Convert to watts   i) 0.5 kW
ii) 1.34 kW   iii) 0.285 kW
iv) 0.08 kW   v) 1.057 kW

1.34 Given that 1 A = 1 000 mA
a) Convert to amperes   i) 650 mA
ii) 75 mA   iii) 2 380 mA
iv) 175 mA   v) 400 mA

b) Convert to milliamperes   i) 0.1 A
ii) 0.25 A   iii) 0.87 A
iv) 1.3 A   v) 10.5 A

1.35 Give the answers to the following in kilowatts:
a) 1.5 kW + 600 W
b) 350 W + 2.7 kW + 0.8 kW
c) 3.5 kW − 750 W
d) 1.05 kW − 380 W
e) 2.9 kW + 450 W + 720 kW
f) 1375 W + 0.6 kW + 2 kW
g) 235 W + 1.7 kW − 850 W
h) 0.95 kW + 375 W − 1.2 kW
i) 1.3 kW × 4
j) 250 W × 7

2 Operations with Fractions

2.1 Vulgar Fractions

A bar 1 m long is sheared into five equal pieces. The length of each piece is a fraction of the original length, and is one-fifth of a metre. If the five pieces are placed end-to-end (fig. 2.1), their total length will equal the original length of the bar.

Placing four pieces end-to-end will give four-fifths of the original length, and so on (fig. 2.2).

This type of fraction has a value which is always less than 1 and is called a **vulgar fraction**. The number above the dividing line, called the *numerator*, is always less than the number below the dividing line, called the *denominator*.

To add vulgar fractions which have the same denominator or common denominator, their numerators are added together. For example,

$$\frac{1}{7} + \frac{3}{7} = \frac{1+3}{7} = \frac{4}{7}$$

$$\frac{2}{9} + \frac{5}{9} = \frac{2+5}{9} = \frac{7}{9}$$

$$\frac{5}{13} + \frac{6}{13} = \frac{11}{13}$$

Subtraction takes place in the same way:

$$\frac{11}{13} - \frac{5}{13} = \frac{6}{13}$$

2.2 Improper Fractions

If two or more vulgar fractions are added together, the resulting answer may be greater than 1. For example,

$$\frac{2}{5} + \frac{4}{5} = \frac{6}{5}$$

This type of fraction where the numerator is greater than the denominator is called an **improper fraction**. Thus, $\frac{9}{5}, \frac{17}{8}, \frac{137}{49}$ are improper fractions.

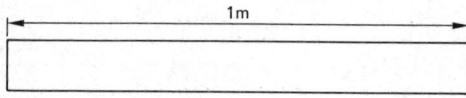

$$\frac{1}{5} + \frac{1}{5} + \frac{1}{5} + \frac{1}{5} + \frac{1}{5} = 1$$

Fig. 2.1

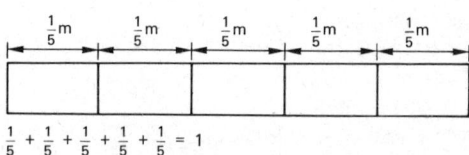

4 PIECES $\frac{1}{5} + \frac{1}{5} + \frac{1}{5} + \frac{1}{5} = \frac{4}{5}$

3 PIECES $\frac{1}{5} + \frac{1}{5} + \frac{1}{5} = \frac{3}{5}$

2 PIECES $\frac{1}{5} + \frac{1}{5} = \frac{2}{5}$

1 PIECE $\frac{1}{5}$

Fig. 2.2

2.3 Mixed Numbers

If the denominator of an improper fraction is divided into the numerator, the resulting answer could be a combination of a whole number and a vulgar fraction, called a **mixed number**. For example,
$$\frac{6}{5} = 1\frac{1}{5}$$
Thus, $1\frac{1}{2}, 3\frac{7}{8}, 5\frac{11}{12}$ are mixed numbers.

Mixed numbers can be converted into improper fractions by multiplying the whole number by the denominator and then adding the numerator.
$$2\frac{3}{5} = \frac{(5 \times 2) + 3}{5} = \frac{13}{5}$$

Example 2.1 Place each of the following values under its correct heading in the table below:

$\frac{3}{8}, 2\frac{1}{4}, \frac{11}{16}, \frac{22}{7}, 4\frac{3}{8}, \frac{11}{10}, 1\frac{7}{9}, \frac{5}{6}, \frac{131}{12}$

VULGAR FRACTION	IMPROPER FRACTION	MIXED NUMBER
$\frac{3}{8}$	$\frac{22}{7}$	$2\frac{1}{4}$
$\frac{11}{16}$	$\frac{11}{10}$	$4\frac{3}{8}$
$\frac{5}{6}$	$\frac{131}{12}$	$1\frac{7}{9}$

2.4 Equivalent Fractions

Fig. 2.3(*a*) shows a circle divided into two equal parts.

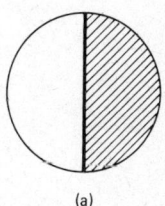 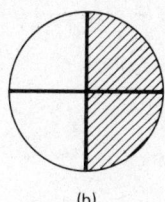

(a)   (b)

Fig. 2.3

The shaded area is equal to one-half of the circle:

$$\text{Shaded area} = \frac{1}{2}$$

Fig. 2.3(*b*) shows the same circle divided into four equal parts. The shaded area is equal to two-fourths of the circle:

$$\text{Shaded area} = \frac{2}{4}$$

The shaded areas in (*a*) and (*b*) are seen to be equal. Therefore the fractions are equal or equivalent:

$$\frac{1}{2} = \frac{2}{4}$$

An **equivalent fraction** for any fraction can be produced by multiplying the numerator and the denominator by the *same number*. For example,

$$\frac{3}{5} = \frac{3 \times 4}{5 \times 4} = \frac{12}{20}$$

$$\therefore \quad \frac{3}{5} = \frac{12}{20}$$

2.5 Cancelling

Equivalent fractions can also be produced for some fractions by *dividing* the numerator and the denominator by the same number. This process is known as **cancelling** and is used to give answers in their lowest terms. For example,

$$\frac{12}{20} = \frac{12 \div 4}{20 \div 4} = \frac{3}{5}$$

$$\therefore \quad \frac{12}{20} = \frac{3}{5}$$

When the numerator and the denominator have large values, it may be necessary to use several stages of cancelling to reach the lowest terms.

$$\frac{126}{378} = \frac{126 \div 9}{378 \div 9} = \frac{14}{42}$$

$$= \frac{14 \div 7}{42 \div 7} = \frac{2}{6}$$

$$= \frac{2 \div 2}{6 \div 2} = \frac{1}{3}$$

$$\frac{180}{240} = \frac{180 \div 10}{240 \div 10} = \frac{18}{24}$$

$$= \frac{18 \div 6}{24 \div 6} = \frac{3}{4}$$

Example 2.2 Cancel each of the following fractions to its lowest terms:

a) $\frac{28}{44}$ *b)* $\frac{12}{9}$ *c)* $\frac{108}{180}$

a) $\frac{28}{44} = \frac{28 \div 4}{44 \div 4} = \frac{7}{11}$ (*Ans.*)

b) $\frac{12}{9} = \frac{12 \div 3}{9 \div 3} = \frac{4}{3} = 1\frac{1}{3}$ (*Ans.*)

c) $\frac{108}{180} = \frac{108 \div 9}{180 \div 9} = \frac{12}{20}$

$$= \frac{12 \div 4}{20 \div 4} = \frac{3}{5}$$ (*Ans.*)

2.6 Addition of Fractions

In order to add fractions they must have the same denominator, i.e. a common denominator. When fractions to be added have different denominators, one or more of them may be replaced by an equivalent fraction to give a common denominator.

Example 2.3 $\dfrac{3}{8} + \dfrac{7}{16}$

Replacing $\frac{3}{8}$ by an equivalent fraction,

$$\frac{3}{8} = \frac{3 \times 2}{8 \times 2} = \frac{6}{16}$$

Now the two fractions have the common denominator 16.

$$\frac{6}{16} + \frac{7}{16} = \frac{13}{16} \quad (Ans.)$$

Example 2.4   Simplify $\dfrac{1}{4} + \dfrac{2}{3}$.

Both fractions must be replaced by equivalent fractions as neither 4 nor 3 can become a common denominator in this case

$$\frac{1}{4} = \frac{1 \times 3}{4 \times 3} = \frac{3}{12}$$

$$\frac{2}{3} = \frac{2 \times 4}{3 \times 4} = \frac{8}{12}$$

The two fractions now have the common denominator 12.

$$\frac{3}{12} + \frac{8}{12} = \frac{11}{12} \quad (Ans.)$$

1   It can be seen from the previous two examples that *the lowest common denominator is the smallest number that both denominators will divide into.* Using this observation the following method may be used to add fractions.

Example $\dfrac{1}{6} + \dfrac{3}{4}$

Step (*a*)   By observation, determine the lowest common denominator of the two fractions.

The lowest common denominator is 12, because it is the smallest number that may be divided by both 6 and 4.

Step (b)   Divide each denominator in turn into the lowest common denominator and multiply the result in each case by the numerator. Then add the products over the lowest common denominator.

Thus $\dfrac{1}{6} + \dfrac{3}{4} = \dfrac{(2 \times 1) + (3 \times 3)}{12}$

$$= \frac{2 + 9}{12}$$

$$= \frac{11}{12} \quad (Ans.)$$

Example 2.5   Find the value of $\frac{5}{6} + \frac{3}{8}$ and give the answer as a mixed number.

Step (*a*)   Lowest common denominator = 24

Step (*b*) $\dfrac{5}{6} + \dfrac{3}{8} = \dfrac{(4 \times 5) + (3 \times 3)}{24}$

$$= \frac{20 + 9}{24}$$

$$= \frac{29}{24} = 1\tfrac{5}{24} \quad (Ans.)$$

2   An alternative method of finding a common denominator is simply to *multiply together the denominators of the fractions to be added.* The resulting answer may be cancelled down to its lowest terms.

Example 2.6   Find the value of $\dfrac{2}{3} + \dfrac{1}{4} + \dfrac{5}{6}$.

Step (*a*)   Common denominator $= 3 \times 4 \times 6 = 72$

Step (*b*) $\dfrac{2}{3} + \dfrac{1}{4} + \dfrac{5}{6} = \dfrac{(24 \times 2) + (18 \times 1) + (12 \times 5)}{72}$

$$= \frac{48 + 18 + 60}{72}$$

$$= \frac{126}{72}$$

$$= \frac{7}{4} \quad \text{(cancelling down)}$$

$$= 1\tfrac{3}{4} \quad (Ans.)$$

Example 2.7   Find the value of $3\frac{1}{2} + 1\frac{3}{4} + 1\frac{5}{8}$.
Two methods may be used to solve this problem.

Method A: bringing the whole numbers together.

$$3\tfrac{1}{2} + 1\tfrac{3}{4} + 1\tfrac{5}{8} = (3 + 1 + 1) + \tfrac{1}{2} + \tfrac{3}{4} + \tfrac{5}{8}$$

$$= 5 + \frac{(4 \times 1) + (2 \times 3) + (1 \times 5)}{8}$$

$$= 5 + \frac{4 + 6 + 5}{8}$$

$$= 5 + \frac{15}{8}$$

$$= 6\tfrac{7}{8} \quad (Ans.)$$

Method B: converting the mixed numbers to improper fractions.

$$3\tfrac{1}{2} + 1\tfrac{3}{4} + 1\tfrac{5}{8} = \frac{7}{2} + \frac{7}{4} + \frac{13}{8}$$

$$= \frac{(4 \times 7) + (2 \times 7) + (1 \times 13)}{8}$$

$$= \frac{28 + 14 + 13}{8}$$

$$= \frac{55}{8} = 6\tfrac{7}{8} \quad (Ans.)$$

2.7 Subtraction of Fractions

Subtraction may also be done by using the methods of equivalent fractions or common denominators.

Example 2.8 Simplify $\tfrac{5}{6} - \tfrac{3}{4}$.

$$\frac{5}{6} - \frac{3}{4} = \frac{(2 \times 5) - (3 \times 3)}{12}$$

$$= \frac{10 - 9}{12} = \frac{1}{12} \quad (Ans.)$$

Example 2.9 Simplify $\tfrac{1}{2} + \tfrac{2}{3} - \tfrac{3}{5}$.

$$\frac{1}{2} + \frac{2}{3} - \frac{3}{5} = \frac{(15 \times 1) + (10 \times 2) - (6 \times 3)}{30}$$

$$= \frac{15 + 20 - 18}{30} = \frac{17}{30} \quad (Ans.)$$

Example 2.10 Simplify $3\tfrac{1}{7} + 2\tfrac{5}{6} - 4\tfrac{3}{14}$.

$$3\tfrac{1}{7} + 2\tfrac{5}{6} - 4\tfrac{3}{14} = (3 + 2 - 4) + \frac{(6 \times 1) + (7 \times 5) - (3 \times 3)}{42}$$

$$= 1 + \frac{6 + 35 - 9}{42}$$

$$= 1 + \frac{32}{42}$$

$$= 1\tfrac{16}{21} \quad (Ans.)$$

2.8 Multiplication of Fractions

Multiplication may be done by multiplying the numerators together and the denominators together. The resulting answer may then be cancelled down to its lowest terms.

Example 2.11 Simplify $\tfrac{3}{8} \times \tfrac{2}{3}$.

$$\frac{3}{8} \times \frac{2}{3} = \frac{3 \times 2}{8 \times 3}$$

$$= \frac{6}{24} = \frac{1}{4} \quad (Ans.)$$

Example 2.12 Simplify $\tfrac{3}{5} \times \tfrac{1}{4} \times \tfrac{2}{3}$.

$$\frac{3}{5} \times \frac{1}{4} \times \frac{2}{3} = \frac{3 \times 1 \times 2}{5 \times 4 \times 3}$$

$$= \frac{6}{60} = \frac{1}{10} \quad (Ans.)$$

Example 2.13 Simplify $1\tfrac{2}{3} \times 3\tfrac{1}{2}$.
First convert the mixed numbers to improper fractions.
$$1\tfrac{2}{3} = \tfrac{5}{3} \qquad 3\tfrac{1}{2} = \tfrac{7}{2}$$

$$\frac{5}{3} \times \frac{7}{2} = \frac{5 \times 7}{3 \times 2}$$

$$= \frac{35}{6} = 5\tfrac{5}{6} \quad (Ans.)$$

2.9 Division of Fractions

When dividing by a fraction, the method used is to invert the divisor and then proceed to multiply. For example
$$\tfrac{3}{4} \div \tfrac{9}{10} \quad \text{becomes}$$

$$\frac{3}{4} \times \frac{10}{9} = \frac{30}{36} = \frac{5}{6}$$

Example 2.14 Simplify $\tfrac{3}{5} \div \tfrac{5}{7}$.

$$\frac{3}{5} \div \frac{5}{7} = \frac{3}{5} \times \frac{7}{5} = \frac{21}{25} \quad (Ans.)$$

Example 2.15 Simplify $\tfrac{4}{5} \div \tfrac{2}{3}$.

$$\frac{4}{5} \div \frac{2}{3} = \frac{4}{5} \times \frac{3}{2}$$

$$= \frac{12}{10} = 1\tfrac{1}{5} \quad (Ans.)$$

Example 2.16 Simplify $1\frac{5}{6} \div 1\frac{1}{4}$.
Convert the mixed numbers to improper fractions.

$$\frac{11}{6} \div \frac{5}{4} = \frac{11}{6} \times \frac{4}{5}$$

$$= \frac{44}{30} = \frac{22}{15} = 1\frac{7}{15} \quad (Ans.)$$

2.10 Fraction of a Quantity

Some workshop problems involve finding a fraction of a quantity.

For example, if $\frac{1}{8}$ of a batch of 320 components are oversize, then the number of oversize components is

$$\frac{1}{8} \times 320 = 40$$

Thus, to find a fraction of a quantity, simply multiply the quantity by the fraction.

Example 2.17 Find
a) $\frac{13}{20}$ of 600 g   b) $\frac{11}{14}$ of 504 cm

a) $\dfrac{13}{20} \times 600 = 390\,\text{g} \quad (Ans.)$

b) $\dfrac{11}{14} \times 504 = 396\,\text{cm} \quad (Ans.)$

Example 2.18 A machine shop reports a total of 48 breakdowns in a year. Two-thirds of these breakdowns were due to electrical faults and the remainder were mechanical failures. Calculate the number of a) electrical faults, b) mechanical failures.

a) Electrical faults $= \frac{2}{3} \times 48 = 32 \quad (Ans.)$

b) Mechanical failures
  = total breakdowns − electrical faults
  $= 48 - 32 = 16 \quad (Ans.)$

Example 2.19 A factory has a workforce of 300 skilled workers: $\frac{11}{20}$ of the workforce are employed on mechanical trades, $\frac{3}{10}$ on electrical trades, and the remainder on fabrication trades.
a) What fraction of the workforce is employed on fabrication trades?
b) Give the number of workers in each trade.

a) The factory has one whole workforce which is divided into three fractions of mechanical, electrical and fabrication trades.

∴ Whole workforce = mech. + elec. + fab.

$$1 = \frac{11}{20} + \frac{3}{10} + \text{fab.}$$

∴ Fab. = whole workforce − mech. − elec.

$$= 1 - \frac{11}{20} - \frac{3}{10}$$

$$= \frac{20 - 11 - 6}{20}$$

$$= \frac{3}{20} \quad (Ans.)$$

b) Number of mechanical workers $= \dfrac{11}{20} \times 300$

$$= 165 \quad (Ans.)$$

Number of electrical workers $= \dfrac{3}{10} \times 300$

$$= 90 \quad (Ans.)$$

Number of fabrication workers $= \dfrac{3}{20} \times 300$

$$= 45 \quad (Ans.)$$

2.11 Expressing One Quantity as a Fraction of Another

If one component in every ten produced in an operation is oversize, then the fraction of oversize components is $\frac{1}{10}$.

Thus, to express one quantity as a fraction of another, that quantity is made the numerator of a fraction where the other quantity is the denominator. The fraction is then cancelled down to its lowest terms.

Example Express 80 as a fraction of 196.
80 becomes the numerator.
196 becomes the denominator.

$$\frac{80}{196} = \frac{20}{49}$$

Example 2.20 In each of the following, state the first quantity as a fraction of the second quantity:

a) 12 m and 20 m   b) 27 kg and 630 kg   c) 200 mm and 875 mm.

a) $\dfrac{12}{20} = \dfrac{3}{5} \quad (Ans.)$

b) $\dfrac{27}{630} = \dfrac{3}{70} \quad (Ans.)$

c) $\frac{200}{875} = \frac{8}{35}$ (*Ans.*)

Example 2.21   In a batch of 900 components, 35 are found to be oversize and 15 undersize. What fraction of the batch is
a) oversize   b) undersize   c) correct size?

a) $\frac{35}{900} = \frac{7}{180}$ oversize   (*Ans.*)

b) $\frac{15}{900} = \frac{1}{60}$ undersize   (*Ans.*)

c) Number of correct size components
$= 900 - 35 - 15 = 850$

$\frac{850}{900} = \frac{17}{18}$ (*Ans.*)

Example 2.22   A warehouse is illuminated by the following electric lamps:

160   100 W
120   60 W
40   40 W

What fraction of the total number of lamps are:
a) 100 W   b) 60 W   c) 40 W?
Total number of lamps = 160 + 120 + 40 = 320

a) $\frac{160}{320} = \frac{1}{2}$ (*Ans.*)

b) $\frac{120}{320} = \frac{3}{8}$ (*Ans.*)

c) $\frac{40}{320} = \frac{1}{8}$ (*Ans.*)

2.12 Conversion of Fractions to Decimals

The fraction $\frac{2}{5}$ means $2 \div 5$, and the answer to this division sum can be given in the form of a decimal fraction:

$$5\overline{)2.0}^{\,0.4} \qquad \therefore \quad \tfrac{2}{5} = 0.4$$

Example 2.23   Convert to decimal fractions:
a) $\frac{7}{8}$ b) $4\frac{9}{16}$

a) $\frac{7}{8} = 7 \div 8$

```
      0.875
   8)7.000
      6 4
      ───
        60
        56
       ───
         40
         40
        ───
         ..
```
$\frac{7}{8} = 0.875$ (*Ans.*)

b) $4\frac{9}{16} = 4 + (9 \div 16)$

```
       0.5625
   16)9.0000
      8 0
      ───
      1 00
        96
       ───
         40
         32
        ───
         80
         80
        ───
         ..
```

$4\frac{9}{16} = 4 + 0.5625 = 4.5625$ (*Ans.*)

Example 2.24   Convert $\frac{2}{13}$ to a decimal fraction and give the answer correct to 3 decimal places.

$$\frac{2}{13} = 2 \div 13$$

As 3 decimal places are required in the answer, the division sum must be worked out to 4 decimal places and rounded-off.

```
      0.1538
   13)2.0000
      1 3
      ───
       70
       65
      ───
        50
        39
       ───
       110
       104
       ───
         6
```

0.1538 to 3 d.p. is 0.154

$\frac{2}{13} = 0.154$ to 3 d.p.   (*Ans.*)

2.13 Conversion of Decimals to Fractions

In a decimal fraction, the first figure to the right of the point gives the number of tenths, the second figure gives hundredths, the third gives thousandths, and the fourth gives ten-thousandths. For example, take 0.4578.

Place value	$\frac{1}{10}$ths	$\frac{1}{100}$ths	$\frac{1}{1\,000}$ths	$\frac{1}{10\,000}$ths
0.4	5	7	8

Thus $0.4 = \dfrac{4}{10}$

$0.05 = \dfrac{5}{100}$ $\qquad$ $0.45 = \dfrac{45}{100}$

$0.007 = \dfrac{7}{1\,000}$ $\qquad$ $0.457 = \dfrac{457}{1\,000}$

$0.000\,8 = \dfrac{8}{10\,000}$ $\qquad$ $0.457\,8 = \dfrac{4\,578}{10\,000}$

When converting to a vulgar fraction, the place value of the last figure of the decimal fraction becomes the denominator, and the whole of the decimal portion is the numerator.

Example 2.25 Convert 0.376 to a vulgar fraction. The last figure of the decimal fraction is 6 which has a place value of thousandths.

$\therefore \quad 0.376 = \dfrac{376}{1\,000} = \dfrac{47}{125}$ $\quad$ (*Ans.*)

Example 2.26 Convert 7.254 to a mixed number. The whole number 7 does not change.

$0.254 = \dfrac{254}{1\,000} = \dfrac{127}{500}$

$\therefore \quad 7.254 = 7\frac{127}{500}$ $\quad$ (*Ans.*)

Exercises 2

In the following exercises the answer should be given in its lowest terms where applicable.

2.1 Simplify $\quad$ a) $\frac{4}{7} + \frac{2}{7}$ $\quad$ b) $\frac{5}{13} + \frac{7}{13}$
c) $\frac{13}{21} + \frac{4}{21} + \frac{2}{21}$ $\quad$ d) $\frac{3}{16} + \frac{7}{16} + \frac{5}{16}$
e) $\frac{2}{5} + 1\frac{1}{5}$ $\quad$ f) $\frac{3}{4} + \frac{1}{4} + 1\frac{1}{4}$
g) $1\frac{1}{8} + 3\frac{5}{8} + \frac{7}{8}$ $\quad$ h) $4\frac{7}{16} + 1\frac{9}{16} + 2\frac{3}{16}$

2.2 Simplify $\quad$ a) $\frac{13}{16} - \frac{5}{16}$
b) $1\frac{5}{8} - \frac{7}{8}$ $\quad$ c) $4\frac{15}{21} - 2\frac{7}{21}$
d) $5\frac{1}{3} - 2\frac{2}{3}$ $\quad$ e) $7\frac{9}{16} - 4\frac{13}{16}$

2.3 Change to improper fractions:
a) $1\frac{3}{4}$ $\quad$ b) $2\frac{11}{12}$ $\quad$ c) $5\frac{3}{4}$
d) $4\frac{7}{8}$ $\quad$ e) $10\frac{11}{16}$

2.4 Change to mixed numbers:
a) $\frac{22}{7}$ $\quad$ b) $\frac{15}{4}$ $\quad$ c) $\frac{33}{12}$
d) $\frac{17}{8}$ $\quad$ e) $\frac{131}{9}$

2.5 Cancel each of the following fractions to its lowest terms:
a) $\frac{18}{24}$ $\quad$ b) $\frac{15}{75}$ $\quad$ c) $\frac{24}{60}$ $\quad$ d) $\frac{130}{150}$
e) $\frac{44}{16}$ $\quad$ f) $\frac{196}{240}$ $\quad$ g) $\frac{38}{57}$ $\quad$ h) $\frac{66}{242}$

2.6 Simplify
a) $\frac{1}{8} + \frac{1}{2}$ $\quad$ b) $\frac{5}{8} + \frac{1}{4}$ $\quad$ c) $\frac{9}{16} + \frac{3}{8}$
d) $\frac{7}{10} + \frac{3}{20}$ $\quad$ e) $\frac{1}{4} + \frac{2}{5}$ $\quad$ f) $\frac{3}{7} + \frac{2}{3}$
g) $\frac{1}{3} + \frac{1}{4} + \frac{1}{6}$ $\quad$ h) $\frac{2}{5} + \frac{7}{10} + \frac{3}{4}$
i) $\frac{3}{4} + \frac{1}{2} + \frac{5}{8}$ $\quad$ j) $\frac{5}{12} + \frac{2}{3} + \frac{3}{8}$
k) $\frac{4}{5} + \frac{7}{10} + \frac{1}{3}$ $\quad$ l) $\frac{5}{6} + \frac{2}{3} + \frac{4}{9}$

2.7 Simplify
a) $1\frac{5}{8} + \frac{3}{4}$ $\quad$ b) $2\frac{1}{2} + 1\frac{1}{4}$
c) $1\frac{7}{10} + 2\frac{9}{20}$ $\quad$ d) $4\frac{5}{16} + 1\frac{3}{8}$
e) $1\frac{3}{8} + 2\frac{1}{4} + 5\frac{1}{2}$ $\quad$ f) $2\frac{2}{3} + 1\frac{5}{6} + 3\frac{1}{4}$
g) $7\frac{5}{8} + 3\frac{7}{12} + 1\frac{1}{6}$ $\quad$ h) $4\frac{3}{5} + 9\frac{7}{10} + 3\frac{3}{4}$

2.8 Simplify $\quad$ a) $\frac{3}{4} - \frac{3}{8}$
b) $\frac{7}{8} - \frac{9}{16}$ $\quad$ c) $\frac{9}{10} - \frac{4}{15}$ $\quad$ d) $1\frac{3}{4} - \frac{7}{8}$
e) $2\frac{1}{2} - 1\frac{3}{4}$ $\quad$ f) $4\frac{1}{4} - 2\frac{5}{8}$ $\quad$ g) $3\frac{3}{10} - 1\frac{11}{20}$
h) $1\frac{1}{3} - \frac{5}{6}$ $\quad$ i) $3\frac{7}{12} - 1\frac{1}{4}$ $\quad$ j) $7\frac{4}{9} - 3\frac{2}{3}$

2.9 Simplify $\quad$ a) $\frac{2}{3} \times \frac{3}{4}$
b) $\frac{4}{5} \times \frac{2}{7}$ $\quad$ c) $\frac{4}{9} \times \frac{5}{6}$ $\quad$ d) $\frac{3}{10} \times \frac{4}{5}$
e) $\frac{7}{8} \times \frac{9}{13}$ $\quad$ f) $\frac{1}{4} \times \frac{2}{3} \times \frac{5}{7}$
g) $\frac{3}{16} \times \frac{4}{5} \times \frac{2}{9}$ $\quad$ h) $\frac{4}{7} \times \frac{3}{4} \times \frac{5}{12}$

2.10 Simplify $\quad$ a) $2\frac{3}{4} \times \frac{1}{2}$
b) $1\frac{5}{8} \times \frac{2}{3}$ $\quad$ c) $\frac{9}{16} \times 3\frac{1}{5}$ $\quad$ d) $1\frac{3}{4} \times 2\frac{5}{7}$
e) $\frac{18}{7} \times 1\frac{4}{9}$ $\quad$ f) $3\frac{4}{5} \times \frac{6}{19}$
g) $1\frac{1}{2} \times 1\frac{3}{4} \times \frac{2}{3}$ $\quad$ h) $2\frac{5}{8} \times 1\frac{1}{7} \times 3\frac{1}{2}$

2.11 Simplify
a) $\frac{1}{2} \div \frac{1}{4}$ $\quad$ b) $\frac{3}{8} \div \frac{3}{4}$ $\quad$ c) $\frac{3}{5} \div \frac{4}{7}$
d) $\frac{9}{10} \div \frac{2}{5}$ $\quad$ e) $\frac{7}{8} \div \frac{3}{4}$ $\quad$ f) $\frac{9}{11} \div \frac{2}{3}$
g) $\frac{2}{3} \div \frac{17}{9}$ $\quad$ h) $\frac{5}{6} \div \frac{2}{11}$

2.12 Simplify
a) $1\frac{1}{2} \div \frac{3}{8}$ $\quad$ b) $2\frac{1}{4} \div \frac{1}{2}$
c) $5\frac{5}{8} \div \frac{5}{8}$ $\quad$ d) $3\frac{9}{10} \div \frac{5}{7}$
e) $2\frac{3}{4} \div 1\frac{1}{2}$ $\quad$ f) $4\frac{1}{8} \div 2\frac{1}{2}$
g) $6\frac{3}{10} \div 1\frac{3}{4}$ $\quad$ h) $2\frac{5}{8} \div 3\frac{1}{2}$

2.13 A batch of 200 electrical resistors were checked for quality, and $\frac{3}{20}$ of the batch were failed by inspection. Determine how many of the resistors were found to be satisfactory.

2.14 An inspector when checking the diameters of a batch of 100 shafts found that $\frac{1}{5}$ of the batch were oversize and $\frac{1}{10}$ undersize. Calculate the number of shafts that were a) oversize, b) undersize, c) correct size.

2.15 In a factory employing 500 skilled workers, $\frac{1}{2}$ are machinists, $\frac{2}{5}$ are assembly workers, and the remainder are maintenance workers. Determine
a) the fraction of the skilled workforce engaged on maintenance
b) the number of maintenance workers employed at the factory.

2.16 Calculate a) $\frac{2}{7}$ of 140 m
b) $\frac{3}{5}$ of 200 kg c $\frac{5}{8}$ of 96 cm
d) $\frac{7}{25}$ of 800 g e) $\frac{4}{15}$ of 135 mm

2.17 Calculate the floor area of a rectangular office of length $6\frac{1}{2}$ m and width $3\frac{1}{4}$ m.

2.18 A model of a light aircraft is to be built to a scale of $\frac{3}{50}$ of full size. If the length of the aircraft is 950 cm and its wingspan is 1 100 cm, calculate the length and the wingspan of the model.

2.19 A manufactured item consists by volume of $\frac{2}{9}$ ferrous metal, $\frac{1}{6}$ non-ferrous metal, and the remainder of plastic material. Determine what fraction of the volume of the item is the plastic material.

2.20 A coolant tank which holds 100 litres is completely filled by a solution made up of $\frac{3}{50}$ soluble oil and the remainder water.
a) What fraction of the solution is the water?
b) How many litres of soluble oil are in the tank?

2.21 Calculate the volume of concrete required to cast a rectangular machine foundation $2\frac{3}{4}$ m long, $1\frac{1}{2}$ m wide and $\frac{3}{4}$ m deep.
(Volume = length × width × depth)

2.22 Convert to decimals:
a) $\frac{3}{4}$ b) $1\frac{4}{5}$ c) $\frac{5}{8}$ d) $2\frac{3}{8}$
e) $\frac{9}{16}$ f) $\frac{17}{20}$ g) $8\frac{13}{50}$ h) $\frac{27}{40}$

2.23 Convert to decimals; give the answers correct to 3 decimal places:
a) $\frac{7}{13}$ b) $1\frac{2}{3}$ c) $3\frac{5}{6}$ d) $5\frac{2}{7}$

2.24 Convert to fractions:
a) 0.7 b) 0.88 c) 0.044 d) 0.175
e) 2.48 f) 5.015 g) 7.364 h) 4.105

2.25 In each of the following give the answer as a decimal:
a) $\frac{3}{4} + 0.172$ b) $0.8 - \frac{5}{8}$
c) $1.706 + \frac{3}{20} + 0.098$ d) $2.5076 - \frac{5}{16}$
e) $\frac{7}{10} + \frac{2}{5} + 0.95$ f) $\frac{3}{4} \div 0.025$
g) $1\frac{7}{8} - 0.6255$ h) $\frac{3}{40} + \frac{7}{20} + \frac{4}{5}$

2.26 Convert to watts (1 kW = 1 000 W):
a) $1\frac{1}{2}$ kW b) $2\frac{3}{4}$ kW c) $4\frac{1}{4}$ kW
d) $2\frac{4}{5}$ kW e) $3\frac{2}{3}$ kW

2.27 Find the total power consumed in kilowatts by the following equipment:
Storage heater $2\frac{1}{2}$ kW
Electric lamp   250 W
Vacuum cleaner $\frac{3}{4}$ kW
Electric kettle   750 W
Electric cooker $10\frac{1}{2}$ kW

2.28 Convert to milliamperes (1 A = 1 000 mA):
a) $1\frac{3}{5}$ A b) $5\frac{1}{3}$ A c) $\frac{3}{4}$ A
d) $9\frac{1}{5}$ A e) $2\frac{1}{4}$ A

2.29 Three resistors of $1\frac{1}{2}$, $2\frac{1}{4}$ and $3\frac{3}{4}$ ohms are connected in series. Find the total resistance.

2.30 The equivalent resistance R_p of two resistors R_1 and R_2, connected in parallel, is given by

$$\frac{1}{R_p} = \frac{1}{R_1} + \frac{1}{R_2}$$

Use this expression to complete the following table:

R_1 OHMS	R_2 OHMS	R_p OHMS
3	6	
2	4	
50	200	
10	40	
3	5	

2.31 Fig. 2.4 shows an electrical circuit containing three resistors connected in series. Calculate the total resistance.

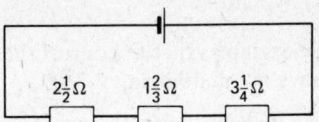

$2\frac{1}{2}\Omega$ $1\frac{2}{3}\Omega$ $3\frac{1}{4}\Omega$

Fig. 2.4

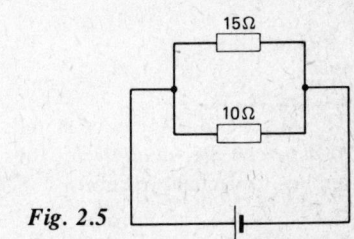

15Ω

10Ω

Fig. 2.5

2.32 Fig. 2.5 shows an electrical circuit containing two resistors connected in parallel. Calculate the equivalent resistance.

3 Percentage, Average and Ratio

3.1 Conversion of Fractions to Percentages

If the fraction of oversize components produced in a machining operation is $\frac{3}{100}$, then 3 components per hundred are oversize. Another term for a hundred is a century.

Thus $\dfrac{3}{100} = 3$ per century

which is abbreviated to

$\dfrac{3}{100} = 3$ per cent   also written as 3%

Any fraction which has the denominator 100 can be written in the same way:

$\dfrac{9}{100} = 9\%$ $\dfrac{31}{100} = 31\%$

Fractions with other denominators can be replaced by equivalent fractions having the denominator 100, to allow easy conversion to percentages. For example,

$$\frac{1}{2} = \frac{1 \times 50}{2 \times 50} = \frac{50}{100} = 50\%$$

$$\frac{3}{4} = \frac{3 \times 25}{4 \times 25} = \frac{75}{100} = 75\%$$

It can be seen from these examples that to convert the fraction to a percentage we are multiplying by 100.

$$\frac{50}{100} \times 100 = 50\%$$

$$\frac{75}{100} \times 100 = 75\%$$

Thus a general rule can be stated: *to convert a fraction to a percentage, multiply by* 100. The same rule also applies to decimal fractions.

For example, to convert $\dfrac{7}{20}$ to a percentage:

$$\frac{7}{20} \times 100 = 35\%$$

and to convert 0.25 to a percentage:

$$0.25 \times 100 = 25\%$$

Example 3.1   Convert to percentages:

a) $\frac{4}{5}$ b) 0.47   c) $\frac{5}{8}$ d) 0.339

a) $\dfrac{4}{5} \times 100 = 80\%$ (*Ans.*)

b) $0.47 \times 100 = 47\%$ (*Ans.*)

c) $\dfrac{5}{8} \times 100 = 62.5\%$ (*Ans.*)

d) $0.339 \times 100 = 33.9\%$ (*Ans.*)

3.2 Conversion of Percentages to Fractions

The process must be reversed to convert a percentage to a fraction. The general rule is: *to convert a percentage to a fraction, divide by* 100.

For example, to convert 42% to a vulgar fraction:

$$42\% = \frac{42}{100} = \frac{21}{50}$$

and to convert 73% to a decimal fraction:

$$73\% = \frac{73}{100} = 0.73$$

Example 3.2   Convert to vulgar fractions:
a) 24%   b) 85%   c) 33.5%

a) $24\% = \dfrac{24}{100} = \dfrac{6}{25}$ (*Ans.*)

b) $85\% = \dfrac{85}{100} = \dfrac{17}{20}$ (*Ans.*)

c) $33.5\% = \dfrac{33.5}{100} = \dfrac{67}{200}$ (*Ans.*)

Example 3.3   Convert to decimal fractions:

a) 14%   b) 68.7%

a) $14\% = \dfrac{14}{100} = 0.14$ *(Ans.)*

b) $68.7\% = \dfrac{68.7}{100} = 0.687$ *(Ans.)*

3.3 Percentage of Quantities

1 The percentage of a quantity can be found by multiplying the quantity by the fraction equivalent of the percentage.

For example, 12% of 400 is

$$12\% = \frac{12}{100} \qquad \frac{12}{100} \times 400 = 48$$

and 35% of 200 m is

$$35\% = \frac{35}{100} \qquad \frac{35}{100} \times 200 = 70\,\text{m}$$

Example 3.4 Find a) 13% of 300 kg
b) 65% of 40 m   c) $22\frac{1}{2}\%$ of 900 l (i.e. 900 litres)

a) $\dfrac{13}{100} \times 300 = 39\,\text{kg}$ *(Ans.)*

b) $\dfrac{65}{100} \times 40 = 26\,\text{m}$ *(Ans.)*

c) $\dfrac{22\frac{1}{2}}{100} \times 900 = 202.5\,\text{l}$ *(Ans.)*

Example 3.5 The capacity of a fuel storage tank is 2 000 l. How much fuel is contained when 45% of the fuel has been used from a full tank?

$$\text{Fuel used} = \frac{45}{100} \times 2\,000 = 900\,\text{l}$$

$$\text{Fuel remaining} = 2\,000 - 900$$
$$= 1\,100\,\text{l} \quad \text{(Ans.)}$$

2 One quantity may be expressed as a percentage of another by showing the relationship as a fraction and then converting to a percentage.

Example 3.6 Give 36 as a percentage of 48.

$$\frac{36}{48} \times 100 = 75\% \quad \text{(Ans.)}$$

Example 3.7 In a batch of 300 resistors 45 are found to be faulty. What percentage of the batch is faulty?

$$\frac{45}{300} \times 100 = 15\% \quad \text{(Ans.)}$$

Example 3.8 In a factory's labour force of 500 people there are 25 who hold first-aid certificates. What percentage of the labour force is qualified in first-aid?

$$\frac{25}{500} \times 100 = 5\% \quad \text{(Ans.)}$$

3.4 Averages

The **average** or arithmetical **mean** of a set of values is given by

$$\text{Average (or mean)} = \frac{\text{sum of the values}}{\text{number of values in the set}}$$

For example, the average of 4, 10, 17, 8, 21 is

$$\text{Average} = \frac{4 + 10 + 17 + 8 + 21}{5} = \frac{60}{5} = 12$$

and the mean of 38, 16, 7, 14, 51, 9, 12 is

$$\text{Mean} = \frac{38 + 16 + 7 + 14 + 51 + 9 + 12}{7}$$

$$= \frac{147}{7} = 21$$

In repetitive machining operations, all the finished components in a batch will never be exactly the same size. Small variations in size will occur due to operational factors such as tool wear, vibration, temperature changes, wear and flexure of the machine parts. However, if the operation is being properly controlled, the variations will be within acceptable limits or tolerance, and the components will not be rejected by inspection.

The average or arithmetical mean size of a number of sample components from a batch is often used to test if a machining operation is under adequate control.

Example 3.9 The measured diameters of a sample of eight shafts produced in a repetitive lathe operation are shown below:

mm 10.08 10.04 10.09 10.14 10.13 10.00 10.22 10.18

Calculate the average diameter of the sample.

Average diameter

$$= \frac{\begin{aligned}&10.08 + 10.04 + 10.09 + 10.14 + \\ &+ 10.13 + 10.00 + 10.22 + 10.18\end{aligned}}{8}$$

$$= \frac{80.88}{8} = 10.11\,\text{mm} \quad \text{(Ans.)}$$

The arithmetic in this example can be made easier by considering only the difference of each size from some chosen dimension, say 10.00 mm. That is

Average variation

$$= \frac{0.08 + 0.04 + 0.09 + 0.14 + \\ + 0.13 + 0.00 + 0.22 + 0.18}{8}$$

$$= \frac{0.88}{8} = 0.11 \text{ mm}$$

Hence   average diameter = 10.00 + 0.11

$$= 10.11 \text{ mm} \quad (Ans.)$$

Example 3.10   The diameter of a machined shaft is measured at five places along its length and the results are shown below:

mm   30.44   30.42   30.55   30.48   30.41

Calculate the mean diameter of the shaft.

Mean diameter

$$= \frac{30.44 + 30.42 + 30.55 + 30.48 + 30.41}{5}$$

$$= \frac{152.30}{5} = 30.46 \text{ mm} \quad (Ans.)$$

Example 3.11   The number of vehicles serviced by a garage over five working days was:

Mon.	Tue.	Wed.	Thu.	Fri.
11	6	9	7	12

Determine the average number of vehicles serviced per day.

$$\text{Average} = \frac{11 + 6 + 9 + 7 + 12}{5} = \frac{45}{5} = 9 \quad (Ans.)$$

Example 3.12   The resistance in ohms of a batch of ten electrical resistors is shown below.

15.1 15.2 14.9 15.0 15.3 15.1 15.1 15.2 15.4 15.3

Calculate the average resistance of the batch.

Average resistance

$$= \frac{15.1 + 15.2 + 14.9 + 15.0 + 15.3 + \\ + 15.1 + 15.1 + 15.2 + 15.4 + 15.3}{10}$$

$$= \frac{151.6}{10} = 15.16 \, \Omega \quad (Ans.)$$

An alternative method is to calculate the average variation from 15 Ω.

Average variation

$$= \frac{0.1 + 0.2 - 0.1 + 0 + 0.3 + \\ + 0.1 + 0.1 + 0.2 + 0.4 + 0.3}{10}$$

$$= \frac{1.6}{10} = 0.16$$

Average resistance = 15 + 0.16 = 15.16 Ω   (Ans.)

Example 3.13   The time taken to assemble each of six gearboxes is shown below.

1 h 35 min	1 h 52 min	1 h 58 min
1 h 31 min	2 h 05 min	1 h 47 min

Calculate the average time required to assemble a gearbox.

To simplify the calculation convert the times to minutes

95   112   118   91   125   107

Average time

$$= \frac{95 + 112 + 118 + 91 + 125 + 107}{6}$$

$$= \frac{648}{6} = 108 \text{ min} = 1 \text{ h } 48 \text{ min} \quad (Ans.)$$

3.5 Ratio

The relationship between two quantities having the same units may be expressed in the form of a **ratio**. For example, consider two metal bars, A and B, where A is 2 m long and B is 3 m long.

The length of A is to the length of B
as   2 is to 3
written   2 : 3

Hence, the ratio of lengths of the two bars A and B is 2 : 3. The ratio shows that for every 2 units of length of A, bar B has 3 units.

Hence $B = A \times \dfrac{3}{2}$ and $A = B \times \dfrac{2}{3}$

Example 3.14   Express as a ratio the speeds of two pulleys, C and D. C makes 30 rev/min and D makes 180 rev/min.

The speed of C is to the speed of D

as   30 is to 180

            30 : 180

             1 : 6   (*Ans.*)

This ratio shows that, for every revolution made by pulley C, pulley D will make six.

Example 3.15   *a*) Find the ratio of masses of two crates A and B. Crate A has a mass of 140 kg and B has a mass of 350 kg.

     Mass of A : mass of B

            140 : 350

              2 : 5   (*Ans.*)

b) If the mass of crate A is increased to 300 kg, what must be the mass of crate B to retain the same ratio of masses?

The ratio 2 : 5 means that, for every 2 kg of A, crate B will have 5 kg.

$$\therefore \quad \text{Mass of B} = \frac{300}{2} \times 5$$

$$= 300 \times \frac{5}{2}$$

$$= 750 \text{ kg}   (\textit{Ans.})$$

Example 3.16   The ratio between the time taken to *replace* a faulty part on a vehicle and the time taken to *repair* the same part is 4 : 5. Determine the time taken to repair the part if it would take 40 min to replace it.

     Replacement time : repair time

                4 : 5

$$\therefore \quad \text{Repair time} = 40 \times \frac{5}{4} = 50 \text{ min}   (\textit{Ans.})$$

Example 3.17   Two metals A and B combine to form an alloy in the ratio of 2 : 7. Determine the mass of metal A required to combine with 56 kg of metal B to form the alloy.

     Mass of A : mass of B

               2 : 7

$$\therefore \quad \text{Mass of A} = \text{mass of B} \times \frac{2}{7}$$

$$= 56 \times \frac{2}{7} = 16 \text{ kg}   (\textit{Ans.})$$

Example 3.18   The composition of a soft solder is given as one part tin and two parts lead.

a) Express the composition as a ratio.

b) What mass of tin is required to alloy with 5 kg of lead?

c) What mass of lead is required to alloy with 2 kg of tin?

a)   Mass of tin : mass of lead

                1 : 2   (*Ans.*)

b)   Mass of tin $= \text{mass of lead} \times \dfrac{1}{2}$

$$= 5 \times \frac{1}{2} = 2\tfrac{1}{2} \text{ kg}   (\textit{Ans.})$$

c) Mass of lead $= \text{mass of tin} \times \dfrac{2}{1}$

$$= 2 \times \frac{2}{1} = 4 \text{ kg}   (\textit{Ans.})$$

Example 3.19   An alloy is made up of three metals A, B and C in the ratio of 2 : 3 : 7. Calculate the amount of each metal in a casting of the alloy having a mass of 36 kg.

The alloy is

   2 parts A + 3 parts B + 7 parts C = 12 parts

Thus   A is 2 parts in $12 = \dfrac{2}{12}$

       B is 3 parts in $12 = \dfrac{3}{12}$

       C is 7 parts in $12 = \dfrac{7}{12}$

$$\therefore \quad \text{Mass of A} = \frac{2}{12} \times 36 = 6 \text{ kg}   (\textit{Ans.})$$

$$\text{Mass of B} = \frac{3}{12} \times 36 = 9 \text{ kg}   (\textit{Ans.})$$

$$\text{Mass of C} = \frac{7}{12} \times 36 = 21 \text{ kg}   (\textit{Ans.})$$

Check   6 + 9 + 21 = 36

Example 3.20   A line of length 900 mm is divided into three sections A, B and C in the ratio 4 : 6 : 10. Determine the length of each section.

   4 + 6 + 10 = 20 parts

$$\text{Length of A} = \frac{4}{20} \times 900 = 180 \text{ mm}   (\textit{Ans.})$$

$$\text{Length of B} = \frac{6}{20} \times 900 = 270 \text{ mm}   (\textit{Ans.})$$

Length of C $= \dfrac{10}{20} \times 900 = 450\,\text{mm}$ (*Ans.*)

Check $180 + 270 + 450 = 900$ mm

Example 3.21   Three electrical resistors R_1, R_2 and R_3 are connected in series and have a total resistance of 48 ohms. Their resistances are in the ratio $3:5:8$. Determine the resistance in ohms of each resistor.

$3 + 5 + 8 = 16$

$R_1 = \dfrac{3}{16} \times 48 = 9\,\Omega$ (*Ans.*)

$R_2 = \dfrac{5}{16} \times 48 = 15\,\Omega$ (*Ans.*)

$R_3 = \dfrac{8}{16} \times 48 = 24\,\Omega$ (*Ans.*)

Check $9 + 15 + 24 = 48$

Exercises 3

3.1 Convert to percentage:
a) $\frac{3}{5}$ b) $\frac{17}{20}$ c) $\frac{9}{25}$
d) $\frac{1}{4}$ e) $\frac{7}{8}$ f) $\frac{11}{16}$

3.2 Convert to percentage:
a) 0.4 b) 0.83 c) 0.17
d) 0.355 e) 0.708 f) 0.015

3.3 Convert to vulgar fractions:
a) 12% b) 64% c) 35%
d) 37.5% e) 81% f) $92\frac{1}{2}$%

3.4 Convert to decimal fractions:
a) 16% b) 74% c) 23.5%
d) $41\frac{1}{2}$% e) 7% f) $80\frac{1}{4}$%

3.5 Determine
a) 25% of 600 mm b) 30% of 200 kg
c) 18% of 4001 d) 60% of 50 kg
e) 87% of 300 cm f) $12\frac{1}{2}$% of 56 m
g) 35.5% of 120 kg h) $22\frac{1}{2}$% of 800 g

3.6 Express
a) 15 as a percentage of 300
b) 24 as a percentage of 60
c) 19 m as a percentage of 200 m
d) 85 kg as a percentage of 125 kg

3.7 A casting has a mass of 40 kg and is produced from an alloy of 88% copper and 12% tin. Determine
a) the fraction of the casting that is copper;
b) the amount of tin required to produce ten of these castings.

3.8 A factory has 120 lathes and wishes to increase its production of turned components by 15%. How many new lathes will be purchased?

3.9 A batch of 500 components produced in a repetitive machining operation is inspected for size. 15 components are found to be undersize and 27 oversize. Determine the percentage of the batch that is a) undersize, b) oversize, c) correct size.

3.10 A component is produced from a cylindrical bar in a lathe operation. The volume of the finished component is 350 cm³ and the amount of metal removed by machining is 150 cm³. Calculate
a) what percentage of the original bar material is in the finished component
b) what percentage of the original material is removed by the cutting operation.

3.11 Calculate the average of
a) 12, 7, 32, 5
b) 22, 74, 10, 41, 63
c) 50.27, 50.11, 50.12, 50.10, 50.18, 50.14, 50.20
d) 0.38, 0.77, 0.65, 0.08, 0.14, 0.44

3.12 Give the mean of
   15.61, 13.85, 14.27, 17.04, 20.09, 16.43
correct to two decimal places.

3.13 The production of a machined component on four working days is shown below:

	Mon.	Tue.	Wed.	Thu.
No. of components	198	264	201	97

Determine the average daily production.

3.14 The test values of a sample of seven electrical resistors are shown below in ohms:
   50.5   51.0   51.5   51.5   49.5   49.0   50.0
Calculate the average resistance of the sample correct to one decimal place.

3.15 The number of absences due to sickness from a large factory was recorded for the first six months of a year and is shown below:

Jan.	Feb.	Mar.	April	May	June
738	621	405	372	393	201

Find the average number of absences per month.

3.16 The results of a tool life test on eight cutting tools are shown below:
   Life in minutes before breakdown
   21   27   34   29   30   28   23   24
Calculate the average tool life.

3.17 The time taken by a motor mechanic to service each of five vehicles is shown below:
   2 h 40 min   3 h 10 min   2 h 50 min   3 h 30 min
   2 h 10 min
Determine the average time taken to service a vehicle.

3.18 Measurements taken of the width of a milled keyway at four positions along its length are shown below:

   mm   12.02   12.05   11.97   12.04

Calculate the average width of the keyway.

3.19 Write out the following ratios in their simplest form:
a) 10:40   b) 12:36   c) 14:91
d) 45:120   e) 85:340   f) 2:4:10
g) 6:12:24   h) 12:21:39

3.20 Express each of the following as a ratio in its simplest form:
a) 8 mm and 72 mm   b) 350 g and 400 g
c) 33 kg and 57 kg   d) 8 m, 20 m and 28 m

3.21 Find the ratio of masses of two castings A and B. Casting A has a mass of 90 kg and B a mass of 162 kg.

3.22 The production of a component requires the following machining times: shaping 35 min, milling 45 min.
a) Give the ratio of shaping to milling times.
b) Calculate the milling time, in the same ratio, which would be required if the shaping time was increased to 42 min.

3.23 Two metals A and B combine to form an alloy in the ratio 8:13. Determine what amount of metal A is required to alloy with 52 kg of metal B.

3.24 A coolant mixture is made up of soluble oil and water in the ratio 1:30. Determine
a) the amount of soluble oil that must be added to 45 litres of water
b) how much water is required to produce 124 litres of the coolant.

3.25 An alloy consists of zinc and copper in the ratio 3:7. Calculate
a) the quantity of zinc required to alloy with 28 kg of copper
b) the quantity of copper required to alloy with 15 kg of zinc
c) the quantity of zinc in an alloy casting of mass 25 kg.

3.26 An alloy consists of three metals A, B and C in the ratio 2:3:5. Find the amount of each metal in an alloy casting of mass 48 kg.

3.27 Concrete for a foundation is made up of 2 parts of cement, 5 parts of sand and 7 parts of stones.
a) Express the mixture in the form of a ratio.
b) Find the quantity required of each constituent to mix 560 kg of concrete.

3.28 The time taken to produce an assembly by machining, fitting and inspection is in the ratio 7:4:1. Calculate the time spent on each process if the total time taken is 3 hours.

3.29 A model is made of an aircraft with dimensions in the ratio 2:15. Determine
a) the length of the model if the length of the aircraft is 30 m
b) the wingspan of the aircraft if the wingspan of the model is 5 m.

3.30 Three shafts A, B and C in a gearbox rotate at speeds in the ratio 1:1.2:1.8. If the speed of shaft A is 550 rev/min, calculate the speeds of the other two shafts.

3.31 A batch of resistors having a nominal value of 250 ohms must be correct within a tolerance of ±10%. What are the maximum and minimum values of resistance for a resistor to be within the tolerance?

3.32 A voltmeter is stated to have an accuracy of ±2%. When the voltmeter gives a reading of 80 V give the maximum and minimum possible values of the measured voltage.

3.33 The *voltage gain* of an amplifier is given by

$$\text{Voltage gain} = \frac{\text{output voltage}}{\text{input voltage}}$$

Use this expression to complete the table

INPUT VOLTAGE	OUTPUT VOLTAGE	VOLTAGE GAIN
3	48	
$\frac{1}{2}$	10	
0.4	15	
$1\frac{1}{2}$	30	
$\frac{4}{5}$	25	

3.34 Three electrical resistors R_1, R_2 and R_3 are connected in series and have a total resistance of 70 ohms. Their resistances are in the ratio 2:5:7. Determine the resistance in ohms of each resistor.

3.35 A workshop store stocks 3 A and 13 A fuses in the ratio 2:7. If the number of 13 A fuses in the store is 322, state how many 3 A fuses are in stock.

4 Use of Mathematical Tables, Calculators and Computers

4.1 Mathematical Tables

Various mathematical tables are available as aids to calculation in workshop problems. They are designed to save effort and time in routine numerical calculations. The tables which are useful at this stage are squares, square roots and logarithms.

Sections of tables are shown in this chapter for the purpose of demonstrating their use, but students will require a set of 4-figure mathematical tables in order to answer the exercises.

It must be emphasised that the simplified treatment given in this chapter is for the benefit of craft students who wish to use mathematical tables as calculating tools. (The treatment is not sufficiently rigorous to be suitable for technician studies; for example, no account is taken of characteristics. However the applications may be useful.)

4.2 Squares of Numbers

1 The **square** of a number is simply the result of multiplying the number by itself.

For example, the square of $9 = 9 \times 9 = 81$

usually written $9^2 = 81$

Example 4.1 Draw up a table of squares for whole numbers from 13 to 20.

Number	Square
13	$13 \times 13 = 169$
14	$14 \times 14 = 196$
15	$15 \times 15 = 225$
16	$16 \times 16 = 256$
17	$17 \times 17 = 289$
18	$18 \times 18 = 324$
19	$19 \times 19 = 361$
20	$20 \times 20 = 400$

Example 4.2 Use the table of squares in Example 4.1 to evaluate:

a) $16^2 + 17^2$ b) $19^2 - 13^2$ c) $15^2 + 14^2 + 18^2$

a) $16^2 + 17^2 = 256 + 289 = 545$ *(Ans.)*
b) $19^2 - 13^2 = 361 - 169 = 192$ *(Ans.)*
c) $15^2 + 14^2 + 18^2 = 225 + 196 + 324$
$\qquad\qquad\qquad = 745$ *(Ans.)*

2 When the number to be squared is large or has a decimal portion, a 4-figure table of squares may be used, a section of which is shown in Table 4.1.

Referring to Table 4.1, the value of 6.2^2 is found on the line of values marked 6.2 and under the column headed 0.

$6.2^2 = 38.44$

The value of 6.25^2 is found on the line marked 6.2 and under the column headed 5.

$6.25^2 = 39.06$

Table 4.1   Squares

	0	1	2	3	4	5	6	7	8	9	Mean Differences								
											1	2	3	4	5	6	7	8	9
6.0	36.00	36.12	36.24	36.36	36.48	36.60	36.72	36.84	36.97	37.09	1	2	4	5	6	7	9	10	11
6.1	37.21	37.33	37.45	37.58	37.70	37.82	37.95	38.07	38.19	38.32	1	2	4	5	6	7	9	10	11
6.2	38.44	38.56	38.69	38.81	38.94	39.06	39.19	39.31	39.44	39.56	1	3	4	5	6	8	9	10	11
6.3	39.69	39.82	39.94	40.07	40.20	40.32	40.45	40.58	40.70	40.83	1	3	4	5	6	8	9	10	11
6.4	40.96	41.09	41.22	41.34	41.47	41.60	41.73	41.86	41.99	42.12	1	3	4	5	6	8	9	10	12

The value of 6.257^2 is found from the value on the line marked 6.2 and under the column headed 5, plus the mean difference value under the column headed 7.

$$6.257^2 = 39.06$$
$$+ \quad 9$$
$$\overline{}$$
$$39.15$$

Note that the place value of the figure in the mean difference column is the same as that of the last figure in the body of the table.

Example 4.3 Use Table 4.1 to evaluate:

a) 6.4^2 b) 6.38^2 c) 6.059^2

a) $6.4^2 \quad = 40.96$ (*Ans.*)
b) $6.38^2 \quad = 40.70$ (*Ans.*)
c) $6.059^2 = 36.60$
$$+ \quad 11$$
$$\overline{}$$
$$36.71 \quad (Ans.)$$

3 Tables of Squares are usually given for values from 1.000 to 9.999, but can be used for any other values by repositioning the decimal point in the answer. This can be done by making a rough estimate of the order or size of the answer before using the tables.

Example 4.4 Find the value of 61.4^2.

Make a rough estimate of the answer by squaring an easy figure which is close to 61.4, say 60

$$60^2 = 60 \times 60 = 3\,600$$

$\therefore \quad 61.4^2$ must be in the order of $3\,600$

Using Table 4.1 and ignoring the place values of the number, the value required is on the line marked 61 and under the column headed 4, giving 37.70.
From the rough estimate the answer is known to be of the order of $3\,600$.

$\therefore \quad 61.4^2 = 3\,770$ (*Ans.*)

Example 4.5 Find 0.6132^2.

Rough estimate $\quad 0.6^2 = 0.6 \times 0.6 = 0.36$

From Table 4.1, 0.6132^2 gives $\quad 37.58$
$$+ \quad 2$$
$$\overline{}$$
$$37.60$$

Repositioning the decimal point to the order of the rough estimate

$$0.6132^2 = 0.3760 \quad (Ans.)$$

4.3 Square Roots of Numbers

1 Finding the square root of a number is the reverse of the process of finding the square. The **square root** of a number is that value which when multiplied by itself equals the number.
For example, the square root of 9 is 3, because

$$3 \times 3 = 9 \quad (\text{i.e.} \quad 3^2 = 9)$$
and is usually written $\quad \sqrt{9} = 3$

In the same way:

$$\sqrt{25} = 5 \quad \text{because } 5^2 = 25$$
$$\sqrt{16} = 4 \quad \text{because } 4^2 = 16$$

Thus the table of squares in Example 4.1 could be rewritten to give the following table of square roots:

Number	Square Root
169	13
196	14
225	15
256	16
289	17
324	18
361	19
400	20

Example 4.6 Use the table of square roots above to evaluate:

a) $\sqrt{196} + \sqrt{324}$ b) $\sqrt{289} \times \sqrt{400}$ c) $17^2 + \sqrt{361}$

a) $\sqrt{196} + \sqrt{324} = 14 + 18 = 32$ (*Ans.*)
b) $\sqrt{289} \times \sqrt{400} = 17 \times 20 = 340$ (*Ans.*)
c) $17^2 + \sqrt{361} = 289 + 19 = 308$ (*Ans.*)

2 Two 4-figure tables of square roots are provided to cover numbers from 1 to 10 and numbers from 10 to 100. Sections of these tables are shown in Table 4.2 and Table 4.3.

The tables of square roots are used in the same manner as the table of squares.

Table 4.2   Square Roots: from 1 to 10

	0	1	2	3	4	5	6	7	8	9	Mean Differences		
											1 2 3	4 5 6	7 8 9
1.5	1.225	1.229	1.233	1.237	1.241	1.245	1.249	1.253	1.257	1.261	0 1 1	2 2 2	3 3 4
1.6	1.265	1.269	1.273	1.277	1.281	1.285	1.288	1.292	1.296	1.300	0 1 1	2 2 2	3 3 3
1.7	1.304	1.308	1.311	1.315	1.319	1.323	1.327	1.330	1.334	1.338	0 1 1	2 2 2	3 3 3
1.8	1.342	1.345	1.349	1.353	1.356	1.360	1.364	1.367	1.371	1.375	0 1 1	1 2 2	3 3 3
1.9	1.378	1.382	1.386	1.389	1.393	1.396	1.400	1.404	1.407	1.411	0 1 1	1 2 2	3 3 3
2.0	1.414	1.418	1.421	1.425	1.428	1.432	1.435	1.439	1.442	1.446	0 1 1	1 2 2	2 3 3

Table 4.3   Square Roots: from 10 to 100

	0	1	2	3	4	5	6	7	8	9	Mean Differences		
											1 2 3	4 5 6	7 8 9
15	3.873	3.886	3.899	3.912	3.924	3.937	3.950	3.962	3.975	3.987	1 3 4	5 6 8	9 10 11
16	4.000	4.012	4.025	4.037	4.050	4.062	4.074	4.087	4.099	4.111	1 2 4	5 6 7	9 10 11
17	4.123	4.135	4.147	4.159	4.171	4.183	4.195	4.207	4.219	4.231	1 2 4	5 6 7	8 10 11
18	4.243	4.254	4.266	4.278	4.290	4.301	4.313	4.324	4.336	4.347	1 2 3	5 6 7	8 9 10
19	4.359	4.370	4.382	4.393	4.405	4.416	4.427	4.438	4.450	4.461	1 2 3	5 6 7	8 9 10
20	4.472	4.483	4.494	4.506	4.517	4.528	4.539	4.550	4.561	4.572	1 2 3	4 6 7	8 9 10

Example 4.7   Find   *a*) $\sqrt{1.964}$ *b*) $\sqrt{18.89}$

a) (Table 4.2) $\sqrt{1.964} = 1.400$
$$+ \quad 1$$
$$\overline{\qquad\qquad}$$
$$1.401 \quad (Ans.)$$

b) (Table 4.3) $\sqrt{18.89} = 4.336$
$$+ \quad 10$$
$$\overline{\qquad\qquad}$$
$$4.346 \quad (Ans.)$$

3   To find the square root of numbers less than 1 or greater than 100, a rough estimate is made to position the decimal point in the answer. This estimate is also used to decide which of the two tables will give the correct value.

Example 4.8   Evaluate $\sqrt{200}$.

Rough estimate: $15^2 = 225$, therefore the answer is of the order of 15.

Table 4.2 gives a value of 1.414, which can be adjusted in accordance with the rough estimate to 14.14 (i.e. in the order of 15).
Table 4.3 gives a value of 4.472, which cannot be adjusted to give a value in the same order as the rough estimate and hence is incorrect.

$\therefore$ $\sqrt{200} = 14.14$ (*Ans.*)

Example 4.9   Evaluate $\sqrt{0.17}$.

Rough estimate: $0.4^2 = 0.16$, therefore the answer is of the order of 0.4.

Table 4.2 gives 1.304 which cannot be adjusted to the order of 0.4 and hence is incorrect.
Table 4.3 give 4.123 which can be adjusted to 0.4123.

$\therefore$ $\sqrt{0.17} = 0.4123$ (*Ans.*)

4.4 Application of Squares and Square Roots

Many workshop problems involve the solution of a **right-angled triangle** to find the length of one side when the lengths of the other two sides are known. For this purpose *Pythagoras' Theorem* is used.

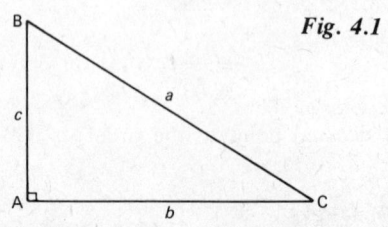

Fig. 4.1

In the triangle shown in fig. 4.1, the angle at A is a right-angle, i.e. angle $A = 90°$. The hypotenuse always lies opposite to the right-angle and is the longest side of the triangle. In this triangle, side a is the hypotenuse and the other two sides are b and c.

The theorem states that *in any right-angled triangle the square on the hypotenuse is equal to the sum of the squares on the other two sides.*

(Length of hypotenuse a)^2
= (length of side b)^2 + (length of side c)^2
$a^2 = b^2 + c^2$

Example 4.10 Calculate the length of side a of the right-angled triangle shown in fig. 4.2.

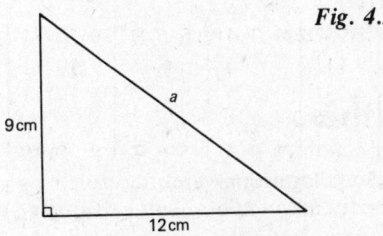

Fig. 4.2

The hypotenuse of the triangle is side a.

$\therefore \quad a^2 = 9^2 + 12^2$
$\quad\quad = 81 + 144 = 225$
$\therefore \quad a = \sqrt{225} = 15$ cm   (*Ans.*)

Example 4.11 Calculate the length of side x in the triangle shown in fig. 4.3.

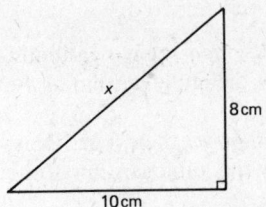

Fig. 4.3

The hypotenuse is side x.

$\therefore \quad x^2 = 10^2 + 8^2$
$\quad\quad = 100 + 64 = 164$
$\therefore \quad x = \sqrt{164}$

Making a rough estimate: $12^2 = 144$, thus x is in the order of 12.
Using square root tables

$x = 12.81$ cm   (*Ans.*)

Example 4.12 Calculate the length of side x in the triangle shown in fig. 4.4.

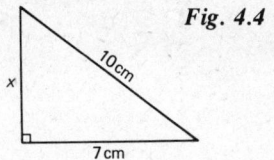

Fig. 4.4

The length of the hypotenuse is 10 cm.

$\therefore \quad 10^2 = 7^2 + x^2$
$\quad\quad 100 = 49 + x^2$
$\quad\quad\quad x^2 = 100 - 49 = 51$
$\quad\quad\quad\quad x = \sqrt{51} = 7.141$ cm   (*Ans.*)

Example 4.13 Determine the centre distance C between the holes in the drilled plate shown in fig. 4.5.

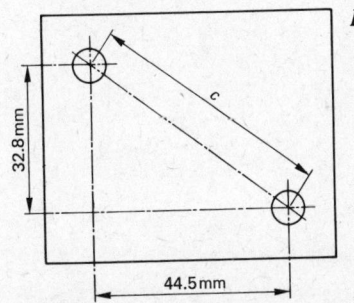

Fig. 4.5

Sketching the triangle (fig. 4.6), the hypotenuse is C.

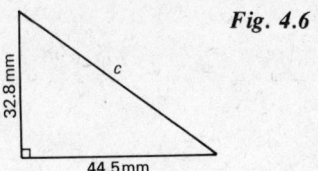

Fig. 4.6

$\therefore \quad C^2 = 44.5^2 + 32.8^2$
$\quad\quad = 1\,980 + 1\,076 = 3\,056$
$\quad\quad C = \sqrt{3\,056} = 55.28$ mm   (*Ans.*)

Example 4.14 Fig. 4.7 shows a 50 mm wide flat milled on an 80 mm diameter bar. Calculate the distance x.

Use of Mathematical Tables, Calculators and Computers 27

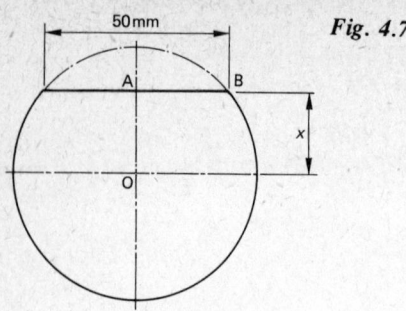

Fig. 4.7

OB is the radius of the bar $= \dfrac{80}{2} = 40\,\text{mm}$

AB is half the width of the flat $= \dfrac{50}{2} = 25\,\text{mm}$

Sketching the triangle (fig. 4.8), the hypotenuse is 40 mm (radius of the bar).

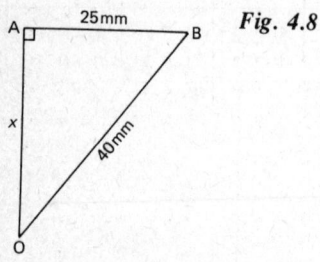

Fig. 4.8

$\therefore\quad 40^2 = 25^2 + x^2$
$1\,600 = 625 + x^2$
$x^2 = 1\,600 - 625 = 975$
$x = \sqrt{975} = 31.22\,\text{mm}\quad (Ans.)$

Example 4.15 The flange shown in fig. 4.9 has four holes drilled on a pitch circle 120 mm diameter. Calculate the centre distance x between the holes A and B.

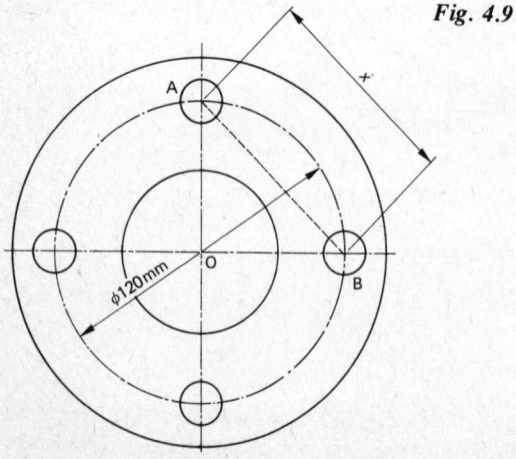

Fig. 4.9

OA and OB are pitch circle radii $= \dfrac{120}{2} = 60\,\text{mm}$

Sketching the triangle (fig. 4.10), the hypotenuse is x.

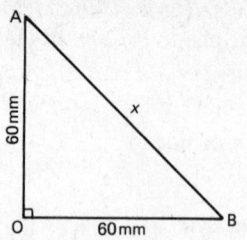

Fig. 4.10

$\therefore\quad x^2 = 60^2 + 60^2$
$= 3\,600 + 3\,600 = 7\,200$
$x = \sqrt{7200} = 84.85\,\text{mm}\quad (Ans.)$

4.5 Logarithms

For the craft student, **logarithms** are used solely as a tool to reduce time spent in tedious multiplication and division. To use logarithms in this simple fashion, it is essential to make a rough estimate of the answer in order to position the decimal point.

Two tables are used:

4-figure logarithms, a section of this table is shown in Table 4.4.
4-figure antilogarithms, a section of this table is shown in Table 4.5.

The tables are read in the same manner as the tables of squares and square roots, adding the mean difference to obtain the fourth-figure.

Two general rules must be remembered:

1 When **multiplying** numbers, *add* their logarithms, and the answer is given by the antilogarithm of the sum.
2 When **dividing** numbers, *subtract* their logarithms, and the answer is given by the antilogarithm of the difference.

Example 4.16 Find the value of 21.58×252.6.
Making a rough estimate of the answer, say $20 \times 250 = 5\,000$. Therefore the answer is of the order of 5 000.
Ignoring the place values of the numbers, their logarithms are found from Table 4.4 and added together.

log of 2158 = 0.3340
log of 2526 = 0.4024
_____
(add)   0.7364

Table 4.4   Logarithms

	0	1	2	3	4	5	6	7	8	9	1 2 3	4 5 6	7 8 9
20	3010	3032	3054	3075	3096	3118	3139	3160	3181	3201	2 4 6	8 11 13	15 17 19
21	3222	3243	3263	3284	3304	3324	3345	3365	3385	3404	2 4 6	8 10 12	14 16 18
22	3424	3444	3464	3483	3502	3522	3541	3560	3579	3598	2 4 6	8 10 12	14 15 17
23	3617	3636	3655	3674	3692	3711	3729	3747	3766	3784	2 4 6	7 9 11	13 15 17
24	3802	3820	3838	3856	3874	3892	3909	3927	3945	3962	2 4 5	7 9 11	12 14 16
25	3979	3997	4014	4031	4048	4065	4082	4099	4116	4133	2 3 5	7 9 10	12 14 15
26	4150	4166	4183	4200	4216	4232	4249	4265	4281	4298	2 3 5	7 8 10	11 13 15
27	4314	4330	4346	4362	4378	4393	4409	4425	4440	4456	2 3 5	6 8 9	11 13 14
28	4472	4487	4502	4518	4533	4548	4564	4579	4594	4609	2 3 5	6 8 9	11 12 14
29	4624	4639	4654	4669	4683	4698	4713	4728	4742	4757	1 3 4	6 7 9	10 12 13
30	4771	4786	4800	4814	4829	4843	4857	4871	4886	4900	1 3 4	6 7 9	10 11 13

Table 4.5   Antilogarithms

	0	1	2	3	4	5	6	7	8	9	1 2 3	4 5 6	7 8 9
.65	4467	4477	4487	4498	4508	4519	4529	4539	4550	4560	1 2 3	4 5 6	7 8 9
.66	4571	4581	4592	4603	4613	4624	4634	4645	4656	4667	1 2 3	4 5 6	7 9 10
.67	4677	4688	4699	4710	4721	4732	4742	4753	4764	4775	1 2 3	4 5 7	8 9 10
.68	4786	4797	4808	4819	4831	4842	4853	4864	4875	4887	1 2 3	4 6 7	8 9 10
.69	4898	4909	4920	4932	4943	4955	4966	4977	4989	5000	1 2 3	5 6 7	8 9 10
.70	5012	5023	5035	5047	5058	5070	5082	5093	5105	5117	1 2 4	5 6 7	8 9 11
.71	5129	5140	5152	5164	5176	5188	5200	5212	5224	5236	1 2 4	5 6 7	8 10 11
.72	5248	5260	5272	5284	5297	5309	5321	5333	5346	5358	1 2 4	5 6 7	9 10 11
.73	5370	5383	5395	5408	5420	5433	5445	5458	5470	5483	1 3 4	5 6 8	9 10 11
.74	5495	5508	5521	5534	5546	5559	5572	5585	5598	5610	1 3 4	5 6 8	9 10 12
.75	5623	5636	5649	5662	5675	5689	5702	5715	5728	5741	1 3 4	5 7 8	9 10 12

The antilogarithm of 0.7364 is found from Table 4.5:

antilog of 0.7364 gives 0.5450

The place value of the answer is now adjusted to agree with the order of the rough estimate.

∴ $21.58 \times 252.6 = 5\,450$ (*Ans.*)

Example 4.17   Evaluate 0.216×0.0241.
Rough estimate: say $0.2 \times 0.02 = 0.004$

(Table 4.4)   log 216 = 0.3345
(Table 4.4)   log 241 = 0.3820
_____
        (add)   0.7165

(Table 4.5)   antilog 0.7165 = 0.5206
∴ $0.216 \times 0.0241 = 0.005\,206$ (*Ans.*)

Example 4.18   Evaluate 5.32×0.64.
Rough estimate: $5 \times 0.6 = 3.0$

Using 4-figure tables:

log 532 = 0.7259
log 64 = 0.8062
_____
(add)   1.5321

The whole number 1 is ignored when finding the antilogarithm.

antilog 0.5321 = 0.3405

∴ $5.32 \times 0.64 = 3.405$ (*Ans.*)

Example 4.19   Find the value of $2.73 \div 1.498$.
Rough estimate: $3 \div 1.5 = 2$

log 273 = 0.4362
log 1498 = 0.1755
_____
(subtract) 0.2607

antilog 0.2607 = 0.1823

∴ $2.73 \div 1.498 = 1.823$ (*Ans.*)

Example 4.20 Evaluate $\dfrac{32.7 \times 12.93}{15.2}$

Rough estimate: $\dfrac{30 \times 13}{15} = 26$

log 327 = 0.5145
log 1293 = 0.1116
_____
(add)   0.6261
log 152 = 0.1818
_____
(subtract) 0.4443

antilog of 0.4443 = 0.2782

$\therefore \dfrac{32.7 \times 12.93}{15.2} = 27.82$ *(Ans.)*

4.6 Use of Calculators

Now that **electronic calculators** are available at relatively low cost, they offer a sensible alternative to logarithms for fast and accurate calculation. Calculators give the answer directly, without the need for rough estimates to position the decimal point.

Students are advised to purchase a simple calculator and become practised at the basic functions of addition, subtraction, multiplication and division. The method of operating the calculator will depend on the model chosen, and the instruction manual should be read carefully. However, the following examples of keyboard operation will apply to most algebraic logic models with an 8-figure display.

Example 4.21 $7.65 + 3.091 + 15.734$

KEYBOARD ENTRY	DISPLAY READS
7.65	7.65
+	7.65
3.091	3.091
+	10.741
15.734	15.734
=	26.475

$7.65 + 3.091 + 15.734 = 26.475$ *(Ans.)*

Example 4.22 $321.76 - 108.309$

KEYBOARD ENTRY	DISPLAY READS
321.76	321.76
−	321.76
108.309	108.309
=	213.451

$321.76 - 108.309 = 213.451$ *(Ans.)*

Example 4.23 43.78×19.67

KEYBOARD ENTRY	DISPLAY READS
43.78	43.78
×	43.78
19.67	19.67
=	861.1526

$43.78 \times 19.67 = 861.1526$ *(Ans.)*

Example 4.24 $13.58 \div 2.46$

KEYBOARD ENTRY	DISPLAY READS
13.58	13.58
÷	13.58
2.46	2.46
=	5.5203252

$13.58 \div 2.46 = 5.5203252$ *(Ans.)*

An answer with such a large number of decimal places would not be convenient or practical to work with in most workshop calculations. Therefore, as shown in Sections 1.6 and 1.7, it would be most useful to give the answer rounded-off to a lesser number of decimal places, e.g.

$13.58 \div 2.46 = 5.52$ correct to 2 d.p.

Example 4.25 $\dfrac{3.95 \times 16.041}{27.42}$ give the answer

to 4 decimal places.

KEYBOARD ENTRY	DISPLAY READS
3.95	3.95
×	3.95
16.041	16.041
÷	63.36195
27.42	27.42
=	2.3107932

$\dfrac{3.95 \times 16.041}{27.42} = 2.3108$ correct to 4 d.p. *(Ans.)*

4.7 Use of a Microcomputer

This section is included to provide an elementary introduction to the use of microcomputers in the solution of simple workshop calculations. All the calculations in this book can be done using a simple hand calculator, but the use of a microcomputer has the advantage where a large number of values are required from a calculation.

Most microcomputers use a programming language which is some version or dialect of BASIC computer language. The programs which follow have been run on the popular Sinclair ZX Spectrum, but may be very easily adapted to suit other models of microcomputer.

PROGRAM 1 This program produces a display of the squares of whole numbers up to 50.

```
10   REM "Squares, 1 TO 50"
20   FOR n = 1 TO 50 STEP 1
30   PRINT n, n * n
40   NEXT n
50   STOP
```

Line 10 commences with the word REM (remark) which informs the computer that the whole of the line can be ignored. The information given after REM is used only to inform the operator of the name or the purpose of the program.

Line 20 includes the commands FOR, TO and STEP which are used to inform the computer of the range of values of n (the number to be squared) over which the program is to be run.

Line 30 tells the computer what calculation must be carried out on the variable n.

Line 40 instructs the computer to move to the next value of n when the squaring of the first value has been carried out. On reaching this instruction, the computer moves back to line 30 and repeats the calculation. This process continues until the stated range of values of n has been covered.

Line 50 completes the program by telling the computer that the program has ended when the last value of n (50) has been reached.

The program is typed into the microcomputer using the keyboard. To run the program the keys RUN and ENTER must be pressed. The DISPLAY on the television screen of the monitor will appear as shown below.

1	1
2	4
3	9
4	16
5	25
6	36
7	49
8	64
9	81
10	100
11	121
12	144
13	169
14	196
15	225
16	256

17	289
18	324
19	361
20	400
21	441
22	484

scroll?

When the computer has filled the screen but has not yet completed the full program, it will stop and the word scroll? will appear at the end of the display. By pressing the ENTER key the operator instructs the computer to display the next screenfull of values. This can occur several times until the whole range of n values has been displayed. The final set of values will show some overlap with preceding values.

Press ENTER

23	529
.	.
.	.
44	1936

scroll?

Press ENTER

29	841
.	.
.	.
50	2500

9 STOP statement, 50:1

PROGRAM 2 This program produces a table showing the conversion of inches to millimetres over the range of 0 to 1 inch in steps of 0.01 inch.

```
10   REM "Metric Conversion"
20   FOR I = 0 TO 1 STEP 0.01
30   PRINT "ins", "mm"
40   PRINT I, 25.4 * I
50   NEXT I
60   STOP
```

Press RUN, press ENTER

ins	mm
.0	0
ins	mm
.01	0.254
ins	mm
.02	0.508
ins	mm
.03	0.762
ins	mm
.04	1.016
ins	mm
.05	1.27
ins	mm
.06	1.524
ins	mm
.07	1.778
ins	mm
.08	2.032
ins	mm
.09	2.286
ins	mm
0.1	2.54

scroll?

Press ENTER

ins	mm
0.11	2.794
.	.
.	.
0.21	5.334

scroll?

The program continues until the whole of the range has been displayed.

PROGRAM 3 This program obtains the area of a circle for any values of radius entered by the operator.

```
10  REM "Area of a circle"
20  PRINT "radius", "area"
30  PRINT
40  INPUT "Enter radius r", r
50  PRINT r, r ↑ 2 * 3.142
60  GO TO 40
```

Line 40: the INPUT command instructs the computer to wait at line 40 while the operator types in any value of radius.

Line 50 instructs the computer to calculate the area using the formula, area $= \pi r^2$. (Note: r^2 is typed as $r \uparrow 2$, and the multiplication symbol is *.)

Line 60 instructs the computer to go back to line 40 to await the next input value of radius.

Press RUN, press ENTER

radius   area

Enter radius r   L   [cursor flashing L]

The operator now types in any value of radius and then presses ENTER. The process can be repeated for as many values as the operator requires. A typical display is shown:

radius	area
10	314.2
20	1256.8
30	2827.8
40	5027.2
50	7855
0.6	1.13112
2.5	19.6375

PROGRAM 4 This program converts any entered value of degrees Fahrenheit to degrees Celsius.

```
10  REM "Convert to Celsius"
20  PRINT "deg F", "deg C"
30  PRINT
40  INPUT "Enter deg F", F
50  PRINT F, (F − 32) * 5/9
60  GO TO 40
```

Line 50 instructs the computer to calculate degrees Celsius using the formula

$$C° = \frac{5}{9}(F° - 32)$$

Press RUN, press ENTER

deg F   **deg C**

Enter deg F ▪L▪

The operator now types in any value of degrees Fahrenheit and then presses ENTER. The process may be repeated any number of times. A typical display is given.

deg F	deg C
40	4.4444444
50	10
60	15.55556
70	21.11111
250	121.11111
975	523.88889
3000	1648.8889

PROGRAM 5 This program converts any entered value of degrees Celsius to degrees Fahrenheit.

```
1Ø   REM "Convert to Fahrenheit"
2Ø   PRINT "deg C", "deg F"
3Ø   PRINT
4Ø   INPUT "Enter deg C", C
5Ø   PRINT C, C * 9/5 + 32
6Ø   GO TO 4Ø
```

Line 5Ø instructs the computer to calculate degrees Fahrenheit using the formula

$$F^\circ = \frac{9}{5} C^\circ + 32$$

Press RUN, press ENTER

deg C   **deg F**

Enter deg C ▪L▪

A typical display is

deg C	deg F
0	32
10	50
20	68
30	86
40	104
50	122
1500	2732

PROGRAM 6 This program obtains the ideal setting of lathe spindle speed in revolutions per minute for any diameter of workpiece when a cutting speed of 25 metres per minute is required.

```
1Ø   REM "Lathe spindle speeds"
2Ø   PRINT "dia", "rev/min"
3Ø   PRINT
4Ø   INPUT "Enter dia d", d
5Ø   PRINT d, (1ØØØ * 25)/(3.142 * d)
6Ø   GO TO 4Ø
```

Line 5Ø instructs the computer to calculate the spindle speed using the formula

$$N = \frac{1\,000\,V}{\pi d}$$

where N = spindle speed (rev/min)
V = cutting speed (m/min)
d = work diameter (mm)

Press RUN, press ENTER

dia   **rev/min**

Enter dia d ▪L▪

The operator now types in any value of work diameter and then presses ENTER. A typical display is given.

dia	rev/min
10	795.67155
20	397.83577
30	265.22385
40	198.91789
50	159.13431
75	106.08954
115	69.18883

PROGRAM 7 This program uses Ohm's Law to find the current flowing in amperes when a resistance in ohms is connected across a 12 volt supply.

```
1Ø    REM "Ohms Law"
2Ø    PRINT "ohms", "amps"
3Ø    PRINT
4Ø    INPUT "Enter ohms R", R
5Ø    PRINT R, 12/R
6Ø    GO TO 4Ø
```

Line 5Ø instructs the computer to calculate the current in amps using Ohms' Law:

$$\text{Amps} = \frac{\text{volts}}{\text{ohms}}$$

Press RUN, press ENTER

ohms   amps

Enter ohms R ▮

The operator now types in any value of resistance and then presses ENTER.

ohms	amps
10	1.2
15	0.8
20	0.6
25	0.48
30	0.4
0.6	20

To practise writing simple programs for workshop calculations and verify them on a microcomputer, the following exercises are suggested:

a) Conversion of gallons to litres.
b) Conversion of yards to metres.
c) Conversion of horsepower to kilowatts.
d) Finding the volume of spheres.
e) Finding supplementary angles using $x + y = 180°$.
f) Finding the total resistance of a series circuit with one resistance varying.
g) Calculating the cutting speed of a milling cutter at various spindle speeds.
h) Calculation of electrical power using $P = EI$ where E or I is constant.

i) Calculation of electrical power using $P = I^2R$ for constant resistance.
j) Calculation of the metal removal rate of a lathe operation at various values of cutting speed.

Exercises 4

4.1 Without using mathematical tables, construct a table of squares for whole numbers from 20 to 30. Use the table to evaluate:
a) $21^2 + 27^2$ b) $29^2 - 26^2$ c) $24^2 + 28^2 - 23^2$

4.2 Use the table from Exercise 4.1 to evaluate
a) $\sqrt{576} + \sqrt{441}$ b) $\sqrt{625} - \sqrt{400}$ c) $27^2 + \sqrt{784}$

4.3 Use 4-figure tables to evaluate:
a) 3.9^2 b) 1.784^2 c) 8.356^2 d) 32.7^2
e) 73.26^2 f) 401.8^2 g) 0.72^2
h) 0.0443^2 i) 0.6482^2 j) $0.031\,59^2$

4.4 Give the answer correct to 2 decimal places:
a) 1.075^2 b) 0.9982^2
c) 5.307^2 d) 0.488^2

4.5 Given that $A = \dfrac{2B^2}{3}$, find the value of A when $B = 6.03$.

4.6 Use 4-figure tables to evaluate:
a) $\sqrt{48}$ b) $\sqrt{2.6}$ c) $\sqrt{120}$ d) $\sqrt{0.62}$
e) $\sqrt{8.349}$ f) $\sqrt{51.73}$ g) $\sqrt{248.6}$
h) $\sqrt{0.3198}$ i) $\sqrt{7432}$ j) $\sqrt{0.019}$

4.7 Give the answer correct to 2 decimal places:
a) $\sqrt{1.943}$ b) $\sqrt{0.876}$
c) $\sqrt{12.44}$ d) $\sqrt{0.079}$

4.8 Give the answer correct to 3 decimal places:
a) $2.177^2 + \sqrt{8.6}$ b) $\sqrt{113} - 1.91^2$
c) $0.634^2 + \sqrt{0.9}$ d) $18.42^2 + \sqrt{18.42}$

4.9 Find the length of the side x in each of the right-angled triangles shown in fig. 4.11.

4.10 Calculate the centre distance D between the bored holes A and B in the bracket shown in fig. 4.12.

4.11 Calculate the centre distance C between the holes in the drilled plate shown in fig. 4.13.

4.12 Calculate the length L on the burned-out plate shown in fig. 4.14; give the answer to the nearest millimetre.

4.13 Calculate the dimension L on the tapered shaft shown in fig. 4.15.

Fig. 4.11

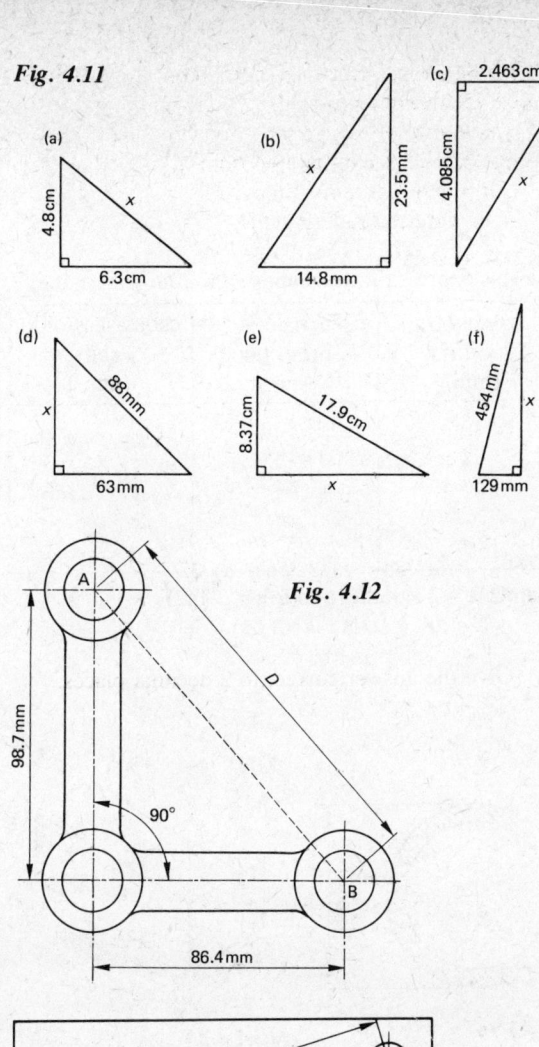

(a)

(b)

(c) 2.463 cm

(d)

(e)

(f)

Fig. 4.12

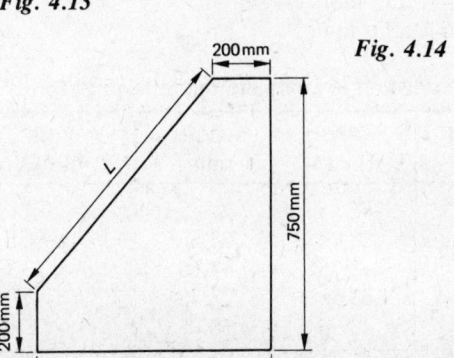

Fig. 4.13

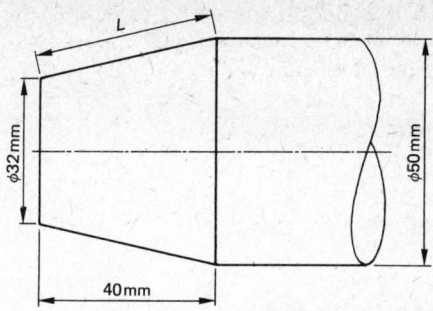

Fig. 4.15

4.14 Determine the length of the diagonal of the square bar shown in fig. 4.16.

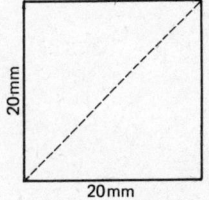

Fig. 4.16

4.15 A rectangular metal plate is 325 mm by 85 mm. Calculate the length of its diagonal.

4.16 Determine the length of the sloping face of the steel wedge shown in fig. 4.17.

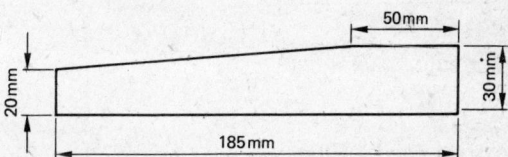

Fig. 4.17

4.17 A casting has a square base of 320 mm side which is to be faced on a centre lathe as shown in fig. 4.18. Determine the distance to the nearest millimetre that the tool must feed in the facing operation.

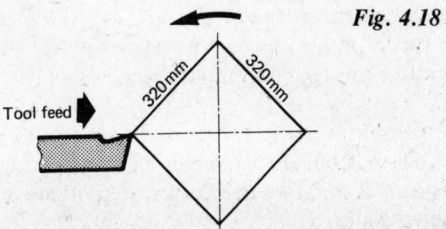

Fig. 4.18

Fig. 4.14

Use of Mathematical Tables, Calculators and Computers 35

Exercises 4.18 to 4.40 may be used for practise in the use of logarithms or calculators; give the answers correct to 4 significant figures.

4.18 1.97×2.34
4.19 4.615×1.203
4.20 19.46×307.5
4.21 143.8×0.175
4.22 0.832×0.767
4.23 $2.918 \times 0.013\,59$
4.24 $7.16 \times 2.245 \times 164.3$
4.25 $18.32 \times 1.218 \times 0.067$
4.26 $3\,156 \times 14.8 \times 0.005\,94$
4.27 $0.318 \times 0.044 \times 0.738$
4.28 $118.6 \div 53.29$
4.29 $1.675 \div 2.074$
4.30 $4\,196 \div 312.5$
4.31 $81.65 \div 0.409$
4.32 $13\,650 \div 9\,157$
4.33 $0.0197 \div 0.002\,45$
4.34 $0.3725 \div 0.000\,94$
4.35 $95.09 \div 1.063$
4.36 $\dfrac{13.38 \times 9.76}{10.44}$
4.37 $\dfrac{4\,216 \times 23.89}{2\,107}$
4.38 $\dfrac{38.4 \times 76.9}{0.737}$
4.39 $\dfrac{0.1695 \times 1.347}{0.0462}$
4.40 $\dfrac{17.05 \times 11.2 \times 9.61}{3\,462}$

Exercises 4.41 to 4.50 are intended for practise in the use of calculators; give the full display answers.

4.41 $7.0631 + 214.9082 + 181.0404$
4.42 $0.6109 + 2.0038 + 0.001\,74 + 19.314$
4.43 $8\,314.96 + 0.004\,19 + 2.070\,31$
4.44 $1\,764.83 - 209.0718$
4.45 $0.076\,385 - 0.009\,174$
4.46 $3.109\,47 - 0.877\,47$
4.47 $2.6014 + 9.3092 - 6.824\,15$
4.48 $47\,621.9 + 171.835 - 1\,010.943$
4.49 $3.86 + 9.42 + 7.07 + 11.23 - 31.58$
4.50 $17.42 + 13.09 + 16.28 + 14.95 + 23.66$

Exercises 4.51 to 4.60 are examples of engineering calculations for which tables of squares, logarithms or calculators may be used.

4.51 The cross-sectional area of a tube (fig. 4.19) may be found from the expression
$$A = \pi(R^2 - r^2)$$
where A = cross-sectional area, mm^2
R = external radius, mm
r = internal radius, mm

Use this expression to complete the following table:

EXTERNAL DIAMETER (mm)	INTERNAL DIAMETER (mm)	CROSS-SECTIONAL AREA (mm²)
40	30	
22	18	
50	44	
32	28	
64	52	

(diameter = 2 × radius; take $\pi = 3.142$).

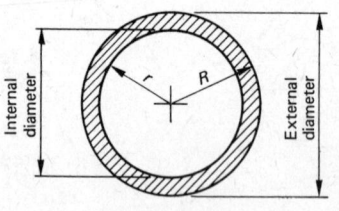

Fig. 4.19   *Fig. 4.20*

4.52 The volume of a rectangular block of metal (fig. 4.20) is given by
$$V = L \times W \times H$$
where V = volume, mm^3
L = length, mm
W = width, mm
H = height, mm

Use this expression to complete the following table:

LENGTH (mm)	WIDTH (mm)	HEIGHT (mm)	VOLUME (mm³)
85	42	27	
53	20.7	25.4	
106.9	31.25	17.68	
244	63.5	28.45	
68.55	43.95	50.05	

(Give the volume correct to 4 significant figures.)

4.53 The cutting speed at the periphery of a bar being turned on a centre lathe (fig. 4.21) is given by

$$V = \frac{\pi dN}{1\,000}$$

where V = cutting speed, m/min
 d = diameter of bar, mm
 N = spindle speed, rev/min

Use this formula to complete the following table:

DIAMETER OF BAR (mm)	SPINDLE SPEED (rev/min)	CUTTING SPEED (m/min)
35	265	
20	550	
12	1 150	
85	115	
15	720	

(Take $\pi = 3.142$ and give the cutting speed correct to 1 decimal place.)

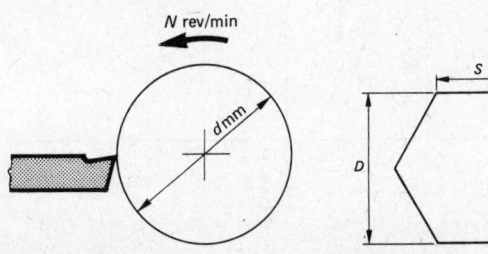

Fig. 4.21   Fig. 4.22

4.54 The distance across the flats of the hexagon shown in fig. 4.22 is given by

$$D = (\sqrt{3}) \times S$$

where D = distance across flats, mm
 S = length of side, mm

Use this formula to complete the following table:

LENGTH OF SIDE (mm)	DISTANCE ACROSS FLATS (mm)
10	
12	
15	
25	
38	

(Give the distance across the flats correct to 2 decimal places.)

4.55 The force required to slide a metal block along a horizontal surface, as shown in fig. 4.23, is given by

$$F = W \times 9.81 \times \mu$$

where F = force required, newtons
 W = mass of block, kilogrammes
 μ = coefficient of friction

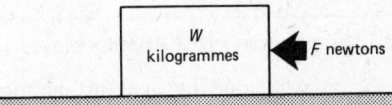

Fig. 4.23

Use the formula to calculate the force required to slide a steel block of mass 13.7 kilogrammes along a horizontal surface when the coefficient of friction between the surfaces is 0.185. Give the answer in newtons correct to 2 decimal places.

4.56 Electrical power is related to current and resistance by the formula

$$P = I^2 R$$

where P = power, watts
 I = current, amps
 R = resistance, ohms

Use this formula to calculate the power loss of a cable having a total resistance of 0.6 ohms and carrying a current of 59.5 amps.

4.57 The formula shown in Exercise 4.56 can be rearranged to give

$$I = \sqrt{\frac{P}{R}}$$

Use this expression to find the value of I when
(i) $P = 160$ watts and $R = 8$ ohms
(ii) $P = 2\,000$ watts and $R = 25$ ohms
(iii) $P = 420$ watts and $R = 16$ ohms
(iv) $P = 1\,800$ watts and $R = 20$ ohms
(v) $P = 800$ watts and $R = 15$ ohms

4.58 The current flowing in an electrical circuit containing two resistors connected in series is given by

$$I = \frac{V}{R_1 + R_2}$$

where V = potential difference, volts
 R_1 and R_2 = resistance, ohms

Use this expression to complete the following table:

R_1 (ohms)	R_2 (ohms)	V (volts)	I (amps)
6	24	12	
8	35	6	
12	30	24	
10	12	7.5	
68	25	110	
18	14	20	

(Give the current in amps correct to 2 decimal places.)

4.59 Calculate the total conducting area of electric cable consisting of 7 wires of 1.35 mm diameter.
(Area of circle $= \pi r^2$; take $\pi = 3.142$.)

4.60 The effective resistance R of a circuit is given by

$$R = R_1 + \left(\frac{R_2 \times R_3}{R_2 + R_3} \right)$$

Determine the value of R when
$R_1 = 10$ ohms
$R_2 = 8$ ohms
$R_3 = 6.5$ ohms

(Give the value of R correct to 2 decimal places.)

4.61 Use the table of Useful Conversion Factors at the front of the book to complete the following (answer to 4 significant figures):

a)	3.746 inches	=	millimetres
b)	8 246 square inches	=	square millimetres
c)	41.6 cubic metres	=	litres
d)	7.304 cubic yards	=	cubic metres
e)	85.6 pounds	=	kilogrammes
f)	416 feet per minute	=	metres per second
g)	1.097 feet	=	metres
h)	54.39 cubic inches	=	cubic millimetres
i)	2.108 yards	=	metres
j)	25.7 miles	=	kilometres
k)	356.3 square yards	=	square metres
l)	844 square feet	=	square metres
m)	216 gallons	=	litres
n)	$3\frac{1}{2}$ horsepower	=	kilowatts

4.62 Use the formulae given below to complete the table (answer to 3 significant figures):

$$I = \frac{E}{R} \qquad P = EI$$

E	R	I	P
20	0.5		
7.5	35		
50	10.5		
12	7		
100	24		

5 Length and Area

5.1 Units of Length

Many countries have now adopted the internationally agreed Système International des Unités or SI System of Units. The basic unit of **length** in the SI system is the **metre**. For technical work, the preferred multiple of the metre is the **kilometre** which equals one thousand metres, and the preferred sub-multiple is the **millimetre** which equals one-thousandth of a metre.

$$1 \text{ km} = 1\,000 \text{ m} \qquad 1 \text{ mm} = \frac{1}{1\,000} \text{ m}$$

In some trades, particularly construction, the **centimetre**, which is equal to one-hundredth of a metre, still has common usage.

$$1 \text{ cm} = \frac{1}{100} \text{ m or } 10 \text{ mm}$$

In decimal form:

```
1 km  = 1 000 m
1 m   =    1 m
1 cm  =    0.01 m
1 mm  =    0.001 m
```

Example 5.1 Convert
a) 2.56 km to metres
b) 5.74 m to millimetres
c) 38 cm to millimetres
d) 819 mm to metres.

a)   1 km = 1 000 m
     2.56 km × 1 000 = 2 560 m   (*Ans.*)

b)   1 m = 1 000 mm
     5.74 m × 1 000 = 5 740 mm   (*Ans.*)

c)   1 cm = 10 mm
     38 cm × 10 = 380 mm   (*Ans.*)

d)   1 000 mm = 1 m
$$\frac{819 \text{ mm}}{1\,000} = 0.819 \text{ m} \quad (Ans.)$$

5.2 Length Conversion Factors

Much existing machinery and equipment has been manufactured in the Imperial or "inch" system of units and, as replacements parts are required, it may be necessary to convert dimensions between the Imperial and the SI systems in order to manufacture the parts on metric machine tools.

Imperial Units		
12 inches	=	1 foot
3 feet	=	1 yard
1 760 yards	=	1 mile

Conversion Factors		
1 inch	=	25.4 mm
1 foot	=	0.3048 m
1 yard	=	0.9144 m
1 mile	=	1.6093 km

(Conversion tables given in BS 350:1959 may also be used.)

Example 5.2 Convert
a) 2.6 inches to millimetres
b) 12 miles to kilometres
c) 88.9 mm to inches
d) 8 yards to metres.

a)   1 inch = 25.4 mm
     2.6 × 25.4 = 66.04 mm   (*Ans.*)

b)   1 mile = 1.6093 km
     12 × 1.6093 = 19.3116 km   (*Ans.*)

c)   25.4 mm = 1 inch
$$\frac{88.9}{25.4} = 3.5 \text{ inches} \quad (Ans.)$$

d)   1 yard = 0.9144 m
     8 × 0.9144 m = 7.3152 m   (*Ans.*)

Example 5.3 Convert
a) 9.2 inches to centimetres
b) 1.4 miles to metres
c) $2\frac{1}{2}$ feet to millimetres

a)   1 inch = 25.4 mm
     9.2 × 25.4 = 233.68 mm

Since 1 cm = 10 mm

$$\frac{233.68\,\text{mm}}{10} = 23.368\,\text{cm} \quad (Ans.)$$

b)   1 mile = 1.6093 km
   1.4 × 1.6093 = 2.253 02 km
Since 1 km = 1 000 m
   2.253 02 km × 1 000 = 2 253.02 m   (Ans.)

c)   1 foot = 0.3048 m
   2.5 × 0.3048 = 0.762 m
Since 1 m = 1 000 mm
   0.762 m × 1 000 = 762 mm   (Ans.)

Example 5.4 Convert to millimetres:
a) 3.194 inches   b) $\frac{5}{8}$ inch   c) $4\frac{3}{8}$ inches
Give the answers correct to 2 decimal places.

a)   1 inch = 25.4 mm
   3.194 × 25.4 = 81.1276 mm
   = 81.13 mm correct to 2 d.p.   (Ans.)

b) $\frac{5}{8}$ × 25.4 = 15.875 mm
   = 15.88 mm correct to 2 d.p.   (Ans.)

c)   4 inches × 25.4 = 101.6 mm
   $\frac{3}{8}$ inch × 25.4 = 9.525 mm

   $4\frac{3}{8}$ inches = 111.125 mm
   = 111.13 mm to 2 d.p.   (Ans.)

Example 5.5 The dimensions of the turned component shown in fig. 5.1 are given in inches. Sketch the component and show the dimensions in millimetres, correct to the nearest 0.01 mm.

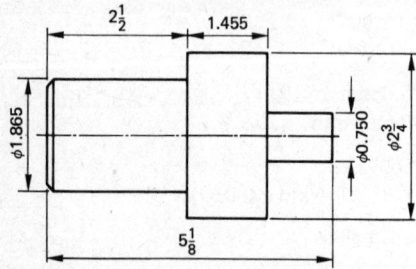

Fig. 5.1

1 inch = 25.4 mm
Conversion of diameters:
   $2\frac{3}{4}$ × 25.4 = 69.85 mm
   1.865 × 25.4 = 47.37 mm
   0.750 × 25.4 = 19.05 mm

Conversion of lengths:
   $5\frac{1}{8}$ × 25.4 = 130.18 mm
   $2\frac{1}{2}$ × 25.4 = 63.50 mm
   1.455 × 25.4 = 36.96 mm

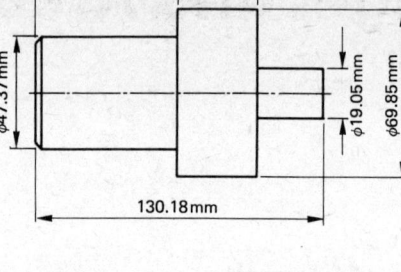

Fig. 5.2

Example 5.6 The tenon on the bored bracket shown in fig. 5.3 is to fit in the machined slot with a clearance of 0.05 mm on both width and depth. Give the dimensions of the tenon in millimetres.

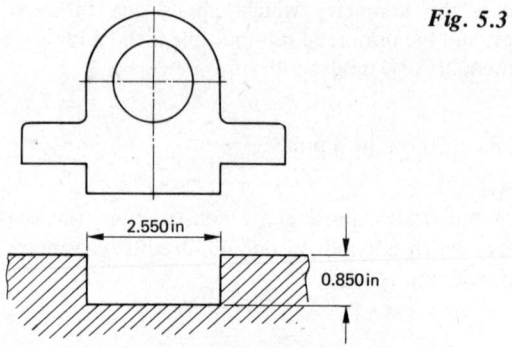

Fig. 5.3

Width of slot = 2.550 × 25.4 = 64.77 mm
Width of tenon = 64.77 − 0.05 = 64.72 mm   (Ans.)
Depth of slot = 0.850 × 25.4 = 21.59 mm
Height of tenon = 21.59 − 0.05 = 21.54 mm   (Ans.)

Example 5.7 In each of the pairs of mating components shown in fig. 5.4, component A is required to fit with component B with the stated clearance between the mating parts. Give the missing dimension of component B in millimetres.

(i)   Diameter of hole in A = 0.950 × 25.4
                              = 24.13 mm
     Missing dimension on B = 24.13 − 0.04
                              = 24.09 mm   (Ans.)

(ii) Width of tenon on A   = 3.050 × 25.4
                              = 77.47 mm
     Missing dimension on B = 77.47 + 0.15
                              = 77.62 mm   (Ans.)

(iii) Width of slot in A   = 0.400 × 25.4
                              = 10.16 mm
     Missing dimension on B = 10.16 − 0.1
                              = 10.06 mm   (Ans.)

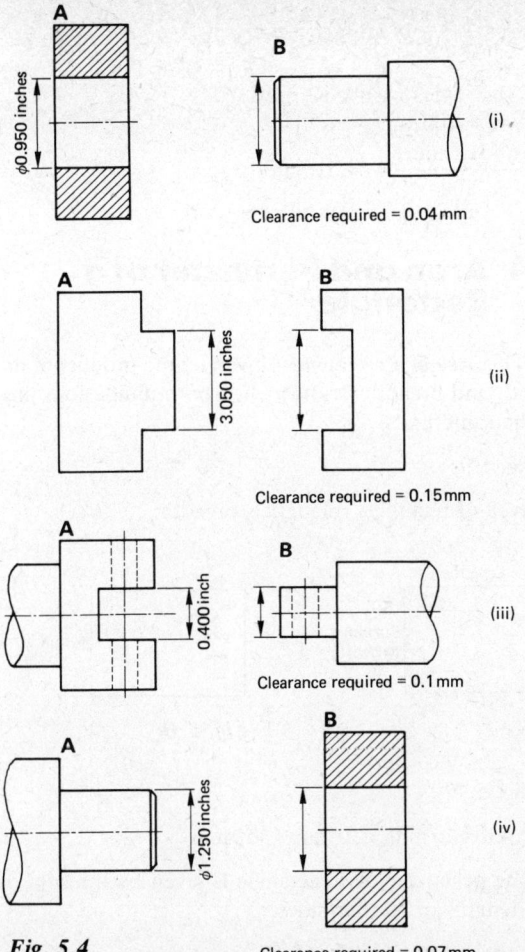

Fig. 5.4

Clearance required = 0.04 mm

Clearance required = 0.15 mm

Clearance required = 0.1 mm

Clearance required = 0.07 mm

(i)
(ii)
(iii)
(iv)

(iv)   Diameter of shaft on A = 1.250×25.4
$\qquad\qquad\qquad\qquad = 31.75\,\text{mm}$
Missing dimension on B $\quad = 31.75 + 0.07$
$\qquad\qquad\qquad\qquad = 31.82\,\text{mm}$ (*Ans.*)

Example 5.8   The design of a guard rail is based on Imperial units and calls for $\frac{1}{2}$ inch diameter bar. The nearest metric size bar available is 12 mm diameter. Express the diameter of the metric bar as a percentage of the design diameter correct to 2 decimal places.

Design dia. $= \frac{1}{2} \times 25.4 = 12.7\,\text{mm}$

$$\text{Percentage} = \frac{\text{dia. of metric bar}}{\text{design dia.}} \times 100$$

$$= \frac{12}{12.7} \times 100$$

$$= 94.49\% \text{ correct to 2 d.p.}   (\textit{Ans.})$$

Example 5.9   A copper fuse element has a diameter of 0.020 inches. What is the equivalent metric size?

$$0.020 \times 25.4 = 0.508\,\text{mm}   (\textit{Ans.})$$

5.3 Units of Area

The **area** of a figure is given by the number of square units that it contains.

1 **square metre** is defined as the area contained by a square of 1 m side.

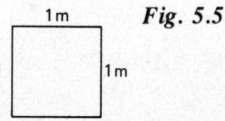

Fig. 5.5

Area $= 1\,\text{m} \times 1\,\text{m} = 1$ square metre
usually written as $1\,\text{m}^2$

In the same way, other units of area are derived from the units of length:

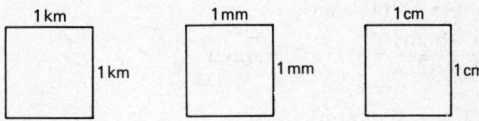

Fig. 5.6

Area $= 1\,\text{km} \times 1\,\text{km} = 1$ square kilometre
$\qquad\qquad\qquad\quad = 1\,\text{km}^2$

Area $= 1\,\text{mm} \times 1\,\text{mm} = 1$ square millimetre
$\qquad\qquad\qquad\quad = 1\,\text{mm}^2$

Area $= 1\,\text{cm} \times 1\,\text{cm} = 1$ square centimetre
$\qquad\qquad\qquad\quad = 1\,\text{cm}^2$

Area may be converted from one unit to another by using a derived conversion factor.

The two squares shown in Fig. 5.7 are exactly the same

Fig. 5.7

Area $= 1\,\text{m} \times 1\,\text{m}$ $\qquad$ Area $= 1\,000\,\text{mm} \times 1\,000\,\text{mm}$
$\quad = 1\,\text{m}^2$ $\qquad\qquad\qquad = 1\,000\,000\,\text{mm}^2$

Hence $1\,\text{m}^2 = 1\,000\,000\,\text{mm}^2$

In the same way the following conversion factors may be derived:

$1 \, km^2 = 1\,000\,000 \, m^2$
$1 \, m^2 = 10\,000 \, cm^2$
$1 \, cm^2 = 100 \, mm^2$

Example 5.10 Convert:
a) $2.5 \, m^2$ to cm^2 b) $3.2 \, km^2$ to m^2
c) $48\,000 \, cm^2$ to m^2 d) $112\,000 \, mm^2$ to cm^2
e) $6\,350\,000 \, mm^2$ to m^2 f) $9\,400\,000 \, m^2$ to km^2

a) $1 \, m^2 = 10\,000 \, cm^2$
   $2.5 \, m^2 \times 10\,000 = 25\,000 \, cm^2$ *(Ans.)*

b) $1 \, km^2 = 1\,000\,000 \, m^2$
   $3.2 \, km^2 \times 1\,000\,000 = 3\,200\,000 \, m^2$ *(Ans.)*

c) $1 \, m^2 = 10\,000 \, cm^2$
   $\dfrac{48\,000 \, cm^2}{10\,000} = 4.8 \, m^2$ *(Ans.)*

d) $1 \, cm^2 = 100 \, mm^2$
   $\dfrac{112\,000 \, mm^2}{100} = 1\,120 \, cm^2$ *(Ans.)*

e) $1 \, m^2 = 1\,000\,000 \, mm^2$
   $\dfrac{6\,350\,000 \, mm^2}{1\,000\,000} = 6.35 \, m^2$ *(Ans.)*

f) $1 \, km^2 = 1\,000\,000 \, m^2$
   $\dfrac{9\,400\,000 \, m^2}{1\,000\,000} = 9.4 \, km^2$ *(Ans.)*

A conversion factor between square inches and square millimetres may be derived by considering two squares of exactly the same size, as shown in fig. 5.8.

Fig. 5.8

   Area $= 1 \, inch \times 1 \, inch$
        $= 1 \, inch^2$

   Area $= 25.4 \, mm \times 25.4 \, mm$
        $= 645.16 \, mm^2$

Hence $1 \, inch^2 = 645.16 \, mm^2$

Example 5.11   Convert:
a) 8.5 square inches to square millimetres
b) 2 000 square millimetres to square inches
c) 10 square inches to square centimetres

a) $1 \, inch^2 = 645.16 \, mm^2$
   $8.5 \, inch^2 \times 645.16 = 5\,483.86 \, mm^2$ *(Ans.)*

b) $\dfrac{2\,000 \, mm^2}{645.16} = 3.1 \, inch^2$ *(Ans.)*

c) $10 \times 645.16 = 6\,451.6 \, mm^2$
    $1 \, cm^2 = 100 \, mm^2$
   $\therefore \quad \dfrac{6\,451.6 \, mm^2}{100} = 64.516 \, cm^2$ *(Ans.)*

5.4 Area and Perimeter of a Rectangle

1   The **area of a rectangle** is given by the product of its length and breadth (width) when both dimensions are in the same units.

In fig. 5.9,

   Area of rectangle $=$ length $\times$ breadth

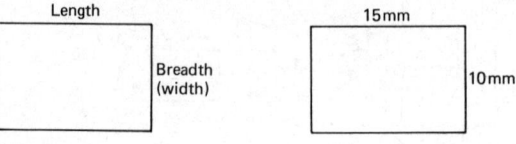

Fig. 5.9   *Fig. 5.10*

In fig. 5.10,

   Area $= 15 \, mm \times 10 \, mm = 150 \, mm^2$

   The **perimeter** of a rectangle is given by the sum of the lengths of its four sides.

For the rectangle in fig. 5.11,

   Length of the rectangle $= a$
   Breadth of the rectangle $= b$
       Perimeter $= a + b + a + b$
               $= 2a + 2b$

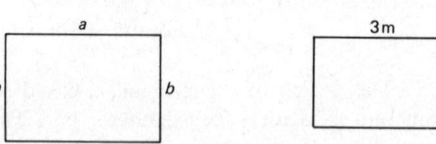

Fig. 5.11   *Fig. 5.12*

In fig. 5.12,

   Perimeter $= (2 \times 3 \, m) + (2 \times 2 \, m)$
            $= 6 \, m + 4 \, m = 10 \, m$

2   The **square** is a rectangle in which all four sides are of equal length. Its area and perimeter are found in the same way as a rectangle.

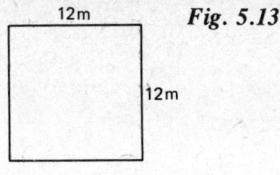

Fig. 5.13

For the square in fig. 5.13,

Area = 12 m × 12 m = 144 m²
Perimeter = 12 + 12 + 12 + 12 = 48 m

Example 5.12 Determine the area and the perimeter of each of the rectangles shown in fig. 5.14

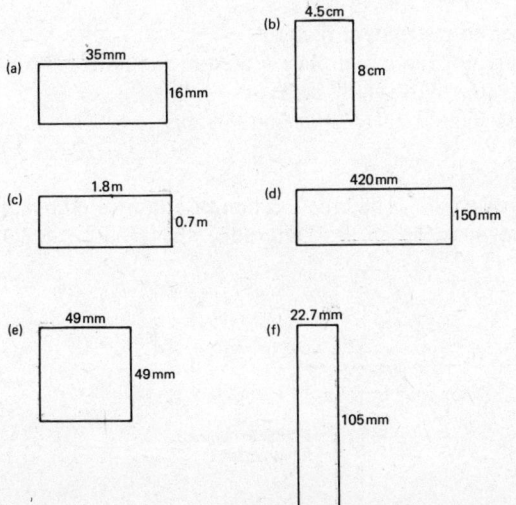

Fig. 5.14

a) Area = 35 mm × 16 mm = 560 mm²
  Perimeter = (2 × 35 mm) + (2 × 16 mm)
    = 70 mm + 32 mm = 102 mm
Area = 560 mm², perimeter = 102 mm   (*Ans.*)

b) Area = 4.5 cm × 8 cm = 36 cm²
  Perimeter = (2 × 4.5 cm) + (2 × 8 cm)
    = 9 cm + 16 cm = 25 cm
Area = 36 cm², perimeter = 25 cm   (*Ans.*)

c) Area = 1.8 m × 0.7 m = 1.26 m²
  Perimeter = (2 × 1.8 m) + (2 × 0.7 m)
    = 3.6 m + 1.4 m = 5 m
Area = 1.26 m², perimeter = 5 m   (*Ans.*)

d) Area = 420 mm × 150 mm = 63 000 mm²
  Perimeter = (2 × 420 mm) + (2 × 150 mm)
    = 840 mm + 300 mm = 1 140 mm
Area = 63 000 mm², perimeter = 1 140 mm   (*Ans.*)

e) Area = 49 mm × 49 mm = 2 401 mm²
  Perimeter = 4 × 49 mm = 196 mm
Area = 2 401 mm², perimeter = 196 mm   (*Ans.*)

f) Area = 22.7 mm × 105 mm = 2 383.5 mm²
  Perimeter = (2 × 22.7 mm) + (2 × 105 mm)
    = 45.4 mm + 210 mm = 255.4 mm
Area = 2 383.5 mm², perimeter = 255.4 mm   (*Ans.*)

3 The dimensions of the rectangle must always be used in the correct length units to give the required units of area; e.g. if the area is required in square millimetres then both dimensions of the rectangle must be in millimetres.

Example 5.13 Calculate the area in square millimetres of each of the rectangles shown in fig. 5.15.

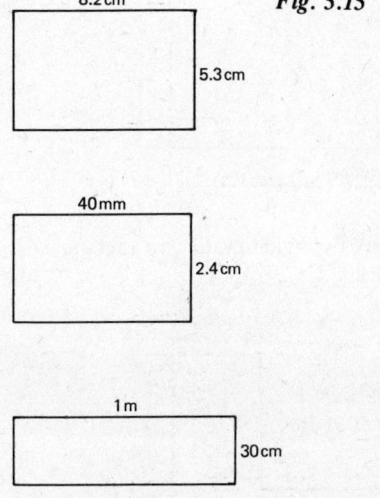

Fig. 5.15

a) Convert both dimensions to millimetres.
Length = 8.2 cm × 10 = 82 mm
Breadth = 5.3 cm × 10 = 53 mm
  Area = 82 × 53 = 4 346 mm²   (*Ans.*)

b) Convert the breadth to millimetres.
Breadth = 2.4 cm × 10 = 24 mm
  Area = 40 × 24 = 960 mm²   (*Ans.*)

c) Convert both dimensions to millimetres.
Length = 1 m × 1 000 = 1 000 mm
Breadth = 30 cm × 10 = 300 mm
  Area = 1 000 × 300 = 300 000 mm²   (*Ans.*)

Example 5.14 Calculate the area of the rectangular face of a cast-iron surface plate, 300 mm by 200 mm.

  Area = length × breadth
    = 300 × 200 = 60 000 mm²   (*Ans.*)

Length and Area 43

Example 5.15 A rectangular inspection panel measures 1 m by 600 mm. Give the area of the panel in square metres.

$$\text{Area} = \text{length} \times \text{breadth}$$
$$= 1\,\text{m} \times \frac{600}{1\,000}\,\text{m}$$
$$= 1 \times 0.6 = 0.6\,\text{m}^2 \quad (Ans.)$$

Example 5.16 Determine the area of the template shown in fig. 5.16.

Fig. 5.16

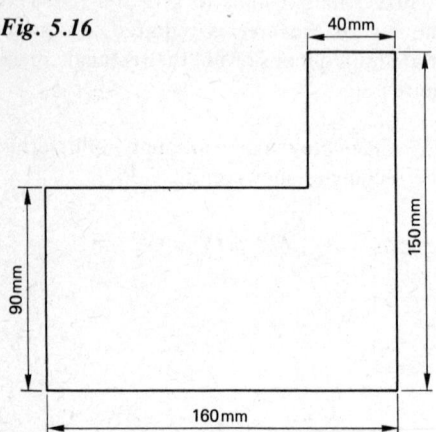

The template may be divided into two rectangles A and B as shown in fig. 5.17.

Fig. 5.17

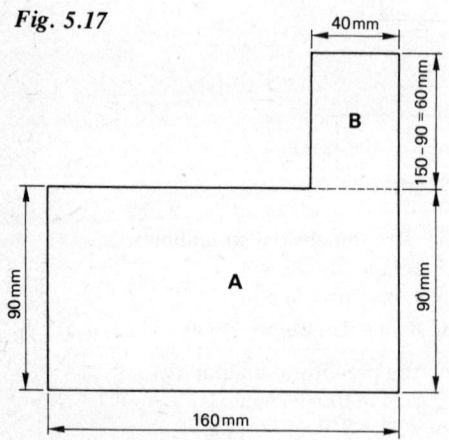

Area of template
= area of rectangle A + area of rectangle B
$= (160 \times 90) + (40 \times 60)$
$= 14\,400 + 2\,400 = 16\,800\,\text{mm}^2 \quad (Ans.)$

Example 5.17 Fig. 5.18 shows an instrument panel which is formed from a metal plate having a rectangular hole punched out to receive the instrument body. Calculate the area of the finished panel.

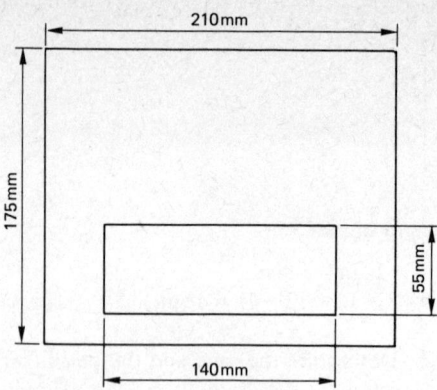

Fig. 5.18

Area of instrument panel
= area of the metal plate − area of the hole
$= (210 \times 175) - (140 \times 55)$
$= 36\,750 - 7\,700 = 29\,050\,\text{mm}^2 \quad (Ans.)$

Example 5.18 The cross-section of a fabricated beam is shown in fig. 5.19. Determine the cross-sectional area.

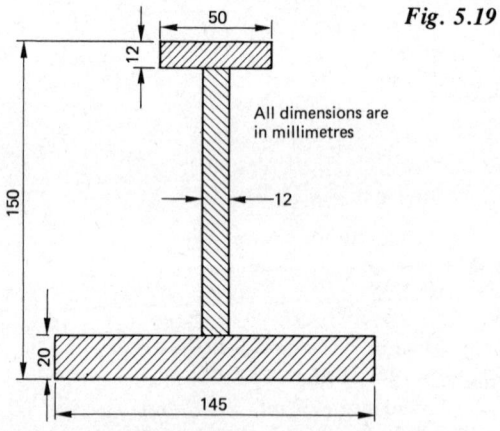

Fig. 5.19

All dimensions are in millimetres

The cross-sectional area of the beam can be divided into three rectangles A, B and C as shown in fig. 5.20.

Cross-sectional area
= area of A + area of B + area of C
$= (145 \times 20) + (12 \times 118) + (50 \times 12)$
$= 2\,900 + 1\,416 + 600$
$= 4\,916\,\text{mm}^2 \quad (Ans.)$

Example 5.19 A rectangular sheet metal cover 260 mm by 240 mm is cut from a plate 300 mm square. Find the remaining area of the plate.

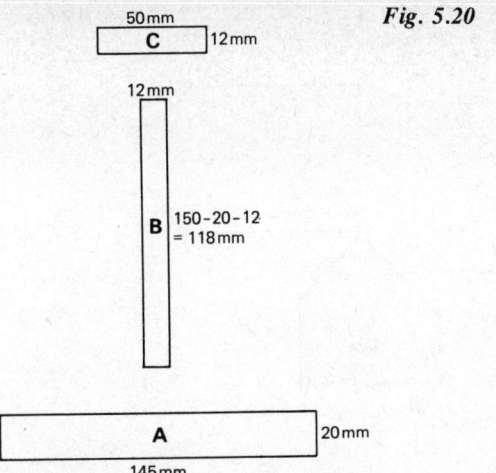

Fig. 5.20

C 50mm 12mm

12mm

B 150−20−12 = 118mm

A 145mm 20mm

Remaining area
= area of the the plate − area of the cover
= (300 × 300) − (260 × 240)
= 90 000 − 62 400 = 27 600 mm²   (*Ans.*)

Example 5.20   Fig. 5.21 shows a sliding operation on a centre lathe. The knife tool is operating with a longitudinal feed of 0.25 mm/rev and a depth of cut of 4.2 mm. Calculate the area of the cut.

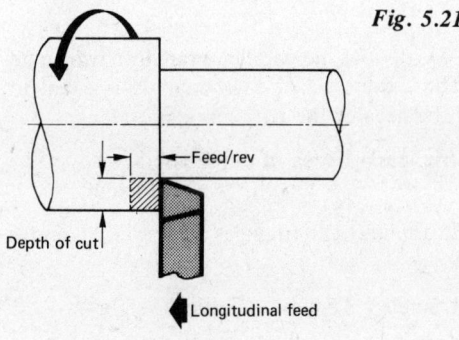

Fig. 5.21

Area of the cut
= area of the shaded rectangle
= feed/rev × depth of cut
= 0.25 × 4.2 = 1.05 mm²   (*Ans.*)

5.5 Area of a Parallelogram

In fig. 5.22, ABCD is a rectangle. If the shaded triangle is cut from one side of the rectangle and attached to the other side, then a new figure ABEF is produced. This figure is a **parallelogram** and is equal in area to the original rectangle.

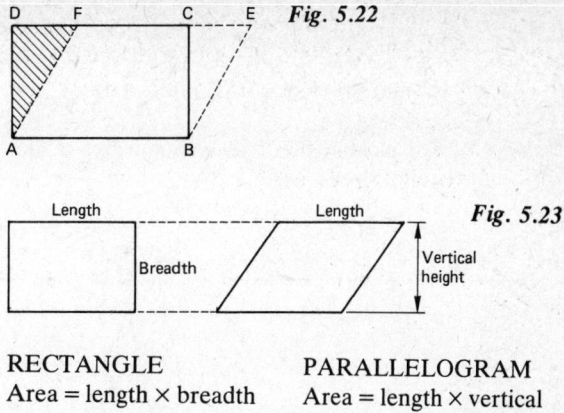

Fig. 5.22

Fig. 5.23

RECTANGLE
Area = length × breadth

PARALLELOGRAM
Area = length × vertical height

Example 5.21   Find the area of the parallelogram shown in fig. 5.24.

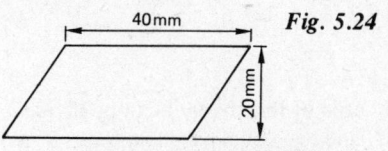

Fig. 5.24

Area = length × vertical height
    = 40 mm × 20 mm = 800 mm²   (*Ans.*)

Example 5.22   Calculate the area in square millimetres of each of the parallelograms shown in fig. 5.25

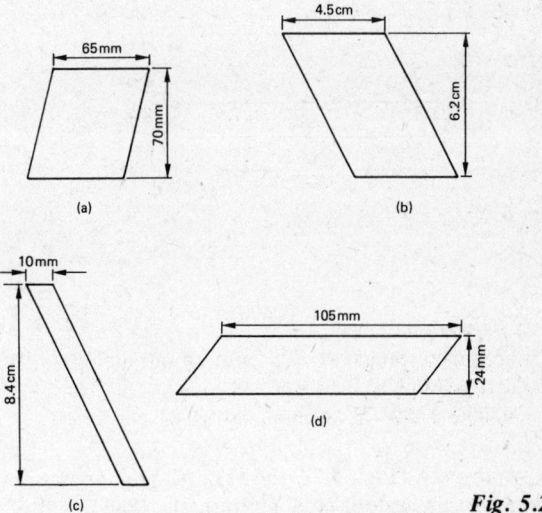

Fig. 5.25

a)   Area = length × vertical height
        = 65 mm × 70 mm
        = 4 550 mm²   (*Ans.*)

b)   Area = 4.5 cm × 6.2 cm
        = 45 mm × 62 mm = 2 790 mm²   (*Ans.*)

c)   Area = 10 mm × 8.4 cm
       = 10 mm × 84 mm = 840 mm² (*Ans.*)

d)   Area = 105 mm × 24 mm = 2 520 mm² (*Ans.*)

Example 5.23 Calculate the cross-sectional area of the fabricated channel shown in fig 5.26.

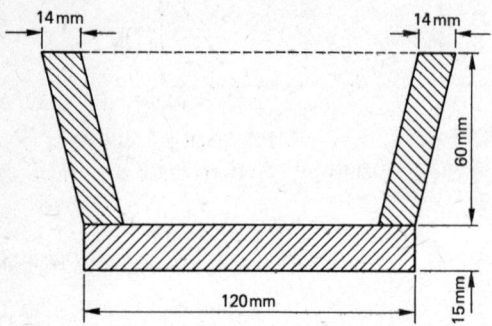

Fig. 5.26

The cross-sectional area of the channel can be divided into the rectangle A and two equal parallelograms B, as shown in fig. 5.27.

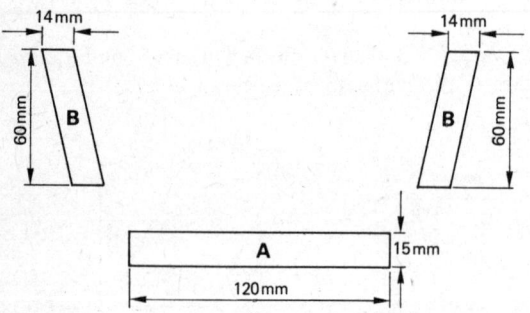

Fig. 5.27

Cross-sectional area
= area of rectangle A + 2 × area of parallelogram B
= (120 × 15) + (2 × 14 × 60)
= 1 800 + 1 680 = 3 480 mm² (*Ans.*)

Example 5.24 Fig. 5.28 shows a sliding operation using a roughing tool on a centre lathe. The depth of cut is 5.8 mm and the longitudinal feed is 0.6 mm/rev. Calculate the area of the cut.

Area of the cut = area of the shaded parallelogram
= feed/rev × depth of cut
= 0.6 × 5.8 = 3.48 mm² (*Ans.*)

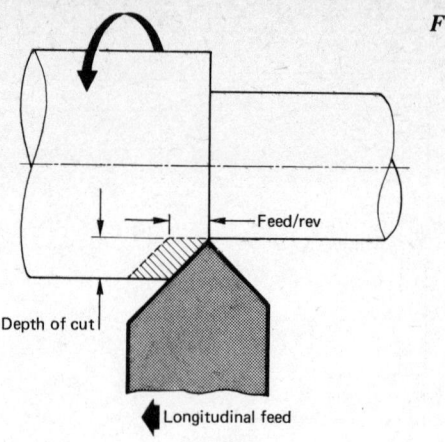

Fig. 5.28

5.6 Area of a Triangle

In fig. 5.29, ABCD is a parallelogram. The diagonal

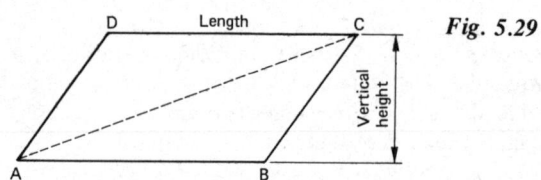

Fig. 5.29

dotted line AC divides the parallelogram into two equal **triangles**. Thus, the area of each triangle is equal to one-half of the area of the parallelogram. Hence,

Area of triangle = ½ area of parallelogram
= ½ × length × vertical height

This formula applies to all triangles (see fig. 5.30) and is usually given as

Area of triangle = ½ × base × vertical height

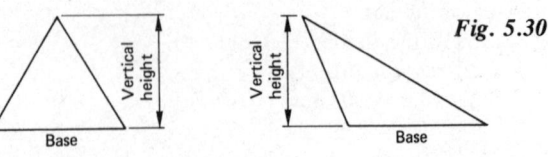

Fig. 5.30

Example 5.25 Find the area of each of the triangles shown in fig. 5.31.

a)   Area of triangle = ½ × base × vertical height
= ½ × 15 mm × 14 mm
= 105 mm² (*Ans.*)

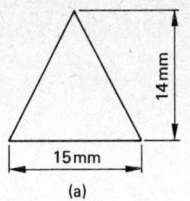

(a)

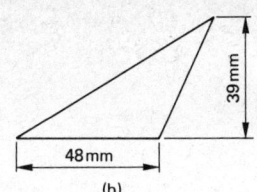

(b)

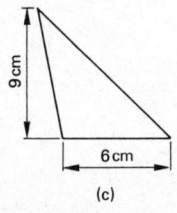

(c)

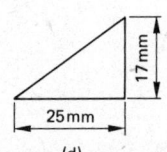

(d)

Fig. 5.31

b) Area $= \frac{1}{2} \times 48\,\text{mm} \times 39\,\text{mm}$
    $= 936\,\text{mm}^2$ (*Ans.*)

c) Area $= \frac{1}{2} \times 6\,\text{cm} \times 9\,\text{cm}$
    $= 27\,\text{cm}^2$ (*Ans.*)

d) Area $= \frac{1}{2} \times 25\,\text{mm} \times 17\,\text{mm}$
    $= 212.5\,\text{mm}^2$ (*Ans.*)

Example 5.26 Determine the area in square millimetres of each of the triangles shown in fig. 5.32.

Fig. 5.32

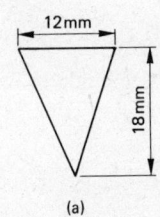

(a)

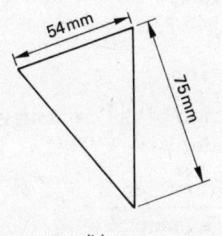

(b)

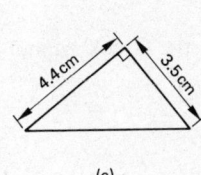

(c)

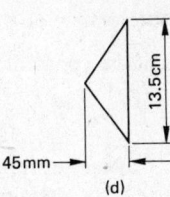

(d)

a) Base $= 12$ mm   vertical height $= 18$ mm
    Area $= \frac{1}{2} \times$ base $\times$ vertical height
    $= \frac{1}{2} \times 12 \times 18 = 108\,\text{mm}^2$ (*Ans.*)

b) Base $= 54$ mm   vertical height $= 75$ mm
    Area $= \frac{1}{2} \times 54 \times 75 = 2\,025\,\text{mm}^2$ (*Ans.*)

c) Base $= 3.5$ cm $= 35$ mm
    Vertical height $= 4.4$ cm $= 44$ mm
    Area $= \frac{1}{2} \times 35 \times 44 = 770\,\text{mm}^2$ (*Ans.*)

d) Base $= 13.5$ cm $= 135$ mm
    Vertical height $= 45$ mm
    Area $= \frac{1}{2} \times 135 \times 45 = 3\,037.5\,\text{mm}$ (*Ans.*)

Example 5.27 Determine the area of a triangle which has a base of 31.5 mm and a vertical height of 164 mm.

Area $= \frac{1}{2} \times$ base $\times$ vertical height
    $= \frac{1}{2} \times 31.5 \times 164 = 2\,583\,\text{mm}^2$ (*Ans.*)

Example 5.28 What area of cut is required to produce the vee-groove in the turned shaft shown in fig 5.33?

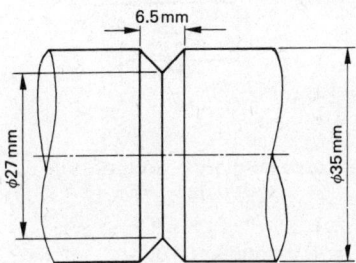

Fig. 5.33

The depth of the vee-groove is the difference between the radii of the shaft at the top and the bottom of the groove (fig. 5.34).

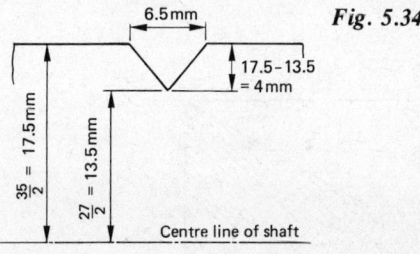

Fig. 5.34

Base $= 6.5$ mm   vertical height $= 4$ mm
    Area $= \frac{1}{2} \times$ base $\times$ vertical height
    $= \frac{1}{2} \times 6.5 \times 4 = 13\,\text{mm}^2$ (*Ans.*)

Example 5.29 Each of the shapes in fig. 5.35 shows a plate of metal which has a hole punched out. Give the remaining area of the plate in each case.

a) Remaining area $=$ area of plate $-$ area of hole
    $= (300 \times 200) - (\frac{1}{2} \times 40 \times 60)$
    $= 60\,000 - 1\,200$
    $= 58\,800\,\text{mm}^2$ (*Ans.*)

b) Remaining area $=$ area of plate $-$ area of hole
    $= (\frac{1}{2} \times 140 \times 180) - (70 \times 30)$
    $= 12\,600 - 2\,100$
    $= 10\,500\,\text{mm}^2$ (*Ans.*)

c) Remaining area $=$ area of plate $-$ area of hole
    $= (105 \times 95) - (\frac{1}{2} \times 32 \times 35)$
    $= 9\,975 - 560$
    $= 9\,415\,\text{mm}^2$ (*Ans.*)

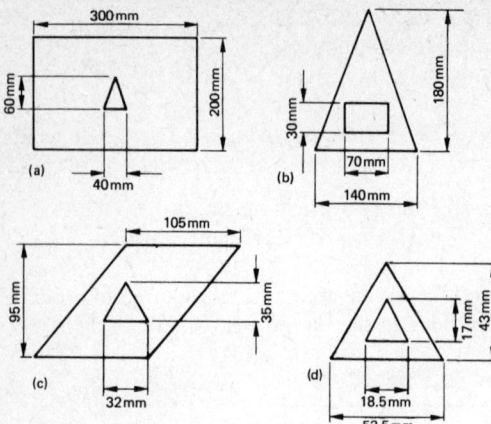

Fig. 5.35

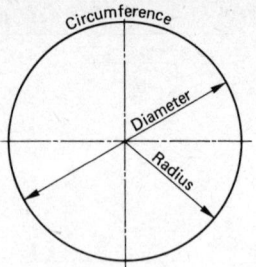

Fig. 5.37

d) Remaining area = area of plate − area of hole
$$= (\tfrac{1}{2} \times 52.5 \times 43) - (\tfrac{1}{2} \times 18.5 \times 17)$$
$$= 1\,128.75 - 157.25$$
$$= 971.5 \text{ mm}^2 \quad (Ans.)$$

Example 5.30 Four triangular pieces A, B, C and D are cut from a metal plate 300 mm by 240 mm as shown in fig. 5.36. The shaded portion is the remaining plate; find its area.

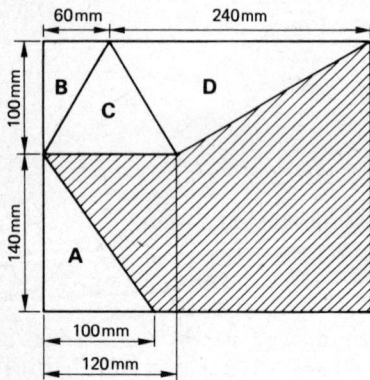

Fig. 5.36

Remaining area = area of plate
\qquad\qquad\qquad − area of the the four triangles
Area of triangle A = $\tfrac{1}{2} \times 100 \times 140 = 7\,000$ mm²
Area of triangle B = $\tfrac{1}{2} \times\;\; 60 \times 100 = 3\,000$ mm²
Area of triangle C = $\tfrac{1}{2} \times 120 \times 100 = 6\,000$ mm²
Area of triangle D = $\tfrac{1}{2} \times 240 \times 100 = 12\,000$ mm²

  Area of the four triangles
  $= 7\,000 + 3\,000 + 6\,000 + 12\,000 = 28\,000$ mm²
  Area of the plate $= 300 \times 240 = 72\,000$ mm²
  Remaining area $= 72\,000 - 28\,000$
  \qquad\qquad\qquad $= 44\,000$ mm² (*Ans.*)

The **Circumference** is the length of the line enclosing the circle (see fig. 5.37).
The **diameter** is the size of the circle measured on a straight line passing through its centre.
The **radius** of the circle is the distance measured on a straight line from its centre to its circumference and is one-half of the diameter.

Let area of the circle = A
\qquad circumference \quad = C
\qquad diameter \qquad\quad = d
\qquad radius \qquad\qquad = r
then

$$d = 2r$$

The ratio of the circumference of any size circle to its diameter is $3\tfrac{1}{7}:1$. That is, for every circle,

$$\frac{\text{Circumference}}{\text{Diameter}} = 3\tfrac{1}{7}$$

∴ Circumference = $3\tfrac{1}{7} \times$ diameter

The value of the constant $3\tfrac{1}{7}$ is given the Greek symbol π (pi).

\qquad $\pi = 3\tfrac{1}{7}$
or \quad $\pi = \tfrac{22}{7}$ \quad as an improper fraction
or \quad $\pi = 3.142$ \quad as a decimal fraction correct to 4 significant figures

Hence,
\quad Circumference = $\pi \times$ diameter
\qquad\qquad\qquad $C = \pi d$
\qquad\qquad or \quad $C = 2\pi r$

The area of a circle is given by

\quad Area = $\pi \times$ radius²
\quad $A = \pi r^2$

Example 5.31 Calculate the circumference and the area of a 14 mm diameter circle.

Diameter = 14 mm
Circumference = $\pi \times$ diameter

$$C = \pi d = \frac{22}{7} \times 14 = 44\,\text{mm} \quad (Ans.)$$

$$\text{Radius} = r = \frac{d}{2} = \frac{14}{2} = 7\,\text{mm}$$

Area = $\pi \times$ radius^2

$$= \pi r^2 = \frac{22}{7} \times 7 \times 7 = 154\,\text{mm}^2 \quad (Ans.)$$

Example 5.32 Calculate the circumference and area of a 28 mm diameter circle.

$$C = \pi d = \frac{22}{7} \times 28 = 88\,\text{mm} \quad (Ans.)$$

$$A = \pi r^2 = \frac{22}{7} \times 14 \times 14 = 616\,\text{mm}^2 \quad (Ans.)$$

Example 5.33 Determine the circumference of each of the following circles:

a) 21 cm diameter
b) 350 mm diameter
c) 84 mm radius

Take $\pi = \dfrac{22}{7}$

a) $\quad C = \pi d = \dfrac{22}{7} \times 21 = 66\,\text{cm} \qquad (Ans.)$

b) $\quad C = \pi d = \dfrac{22}{7} \times 350 = 1100\,\text{mm} \quad (Ans.)$

c) $\quad d = 2r = 2 \times 84 = 168\,\text{mm}$

$$C = \pi d = \frac{22}{7} \times 168 = 528\,\text{mm} \quad (Ans.)$$

Example 5.34 Find the circumference of a circle of 44.5 mm diameter. Take $\pi = 3.142$; give the answer correct to 2 decimal places.

$$\begin{aligned} C &= \pi d \\ &= 3.142 \times 44.5 \\ &= 139.819 = 139.82\,\text{mm correct to 2 d.p.} \quad (Ans.) \end{aligned}$$

Example 5.35 Calculate the area of each of the following circles:

a) 21 mm radius
b) 84 mm diameter
c) 140 cm diameter

Take $\pi = \dfrac{22}{7}$.

a) $\quad A = \pi r^2 = \dfrac{22}{7} \times 21 \times 21 = 1386\,\text{mm}^2 \quad (Ans.)$

b) $\quad r = \dfrac{d}{2} = \dfrac{84}{2} = 42\,\text{mm}$

$$A = \pi r^2 = \frac{22}{7} \times 42 \times 42 = 5544\,\text{mm}^2 \quad (Ans.)$$

c) $\quad r = \dfrac{d}{2} = \dfrac{140}{2} = 70\,\text{cm}$

$$A = \pi r^2 = \frac{22}{7} \times 70 \times 70 = 15\,400\,\text{cm}^2 \quad (Ans.)$$

Example 5.36 Find the area of a circle of 5.7 m diameter. Take $\pi = 3.142$; give the answer correct to 4 significant figures.

$$\text{Radius} = \frac{d}{2} = \frac{5.7}{2} = 2.85\,\text{m}$$

$$\begin{aligned} \text{Area} &= \pi r^2 \\ &= 3.142 \times 2.85 \times 2.85 \\ &= 25.5209 = 25.52\,\text{m}^2 \text{ correct to 4 s.f.} \quad (Ans.) \end{aligned}$$

(Students may find it convenient to use a table of squares to find values such as 2.85^2.)

Example 5.37 Calculate the cross-sectional area of the cast-iron pipe shown in fig. 5.38.

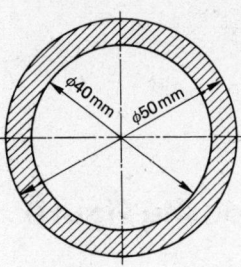

Fig. 5.38

Cross-sectional area of pipe
= area of the outer circle − area of the inner circle
= $(\pi \times 25^2) - (\pi \times 20^2)$
= $\left(\dfrac{22}{7} \times 25 \times 25\right) - \left(\dfrac{22}{7} \times 20 \times 20\right)$
= $1964 - 1257 = 707\,\text{mm}^2 \quad (Ans.)$

Example 5.38 Find the cross-sectional area of the bearing block shown in fig. 5.39.

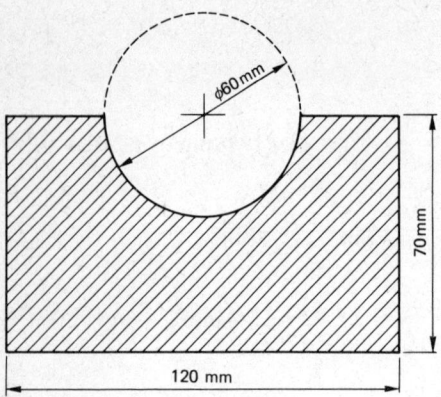

Fig. 5.39

Cross-sectional area
= the shaded area
= area of rectangle – area of semi-circle
= $(120 \times 70) - (\frac{1}{2}\pi r^2)$
= $(120 \times 70) - \left(\frac{1}{2} \times \frac{22}{7} \times 30 \times 30\right)$
= $8\,400 - 1\,414 = 6\,986\,\text{mm}^2$ *(Ans.)*

Example 5.39 Calculate the total conducting area of electric cable consisting of 7 wires of 0.85 mm diameter. (Give the answer correct to 2 decimal places.)

$$\text{Radius of wire} = \frac{0.85}{2} = 0.425\,\text{mm}$$

Area of 1 wire = πr^2
$\qquad = 3.142 \times 0.425 \times 0.425$
$\qquad = 0.5675\,\text{mm}^2$

Total conducting area
$\quad = 7 \times 0.5675$
$\quad = 3.9725\,\text{mm}^2$
$\quad = 3.97\,\text{mm}^2$ correct to 2 d.p.   *(Ans.)*

5.8 Area of a Composite Figure

The area of some complicated shapes can often be found by **dividing the figure into a number of common shapes** such as rectangles, triangles and parts of a circle.

Example 5.40 Calculate the area of the sheet metal template shown in fig. 5.40.

The area of the template can be divided into four common shapes as indicated by the dotted lines:

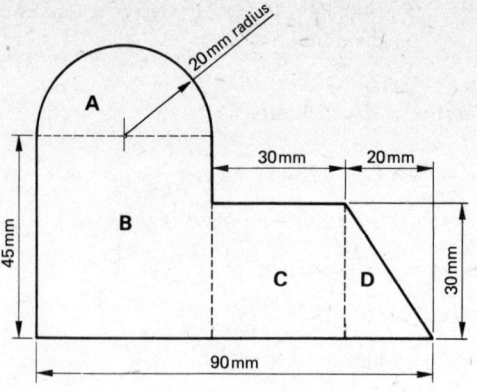

Fig. 5.40

Area of template
= area of semi-circle A + area of rectangle B
$\quad$ + area of square C + area of triangle D
= $\left(\frac{1}{2} \times \frac{22}{7} \times 20 \times 20\right) + \left(40 \times 45\right)$

$\quad + \left(30 \times 30\right) + \left(\frac{1}{2} \times 20 \times 30\right)$

= $628.6 + 1\,800 + 900 + 300$

= $3\,628.6\,\text{mm}^2$ *(Ans.)*

Fig. 5.41 shows some examples of complicated shapes which can be divided by the dotted lines into a number of common shapes in order to find the area.

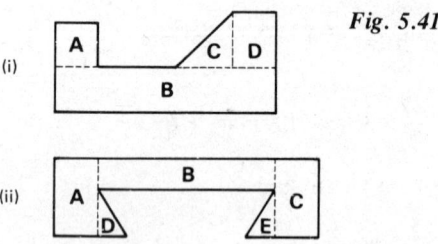

Fig. 5.41

(i)

(ii)

(iii)

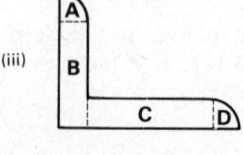

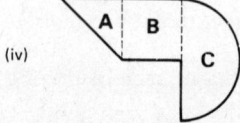

(iv)

Shape (i) = rectangle A + rectangle B
      + triangle C + rectangle D

Shape (ii) = rectangle A + rectangle B + rectangle C
      + triangle D + triangle E

Shape (iii) = $\frac{1}{4}$ circle A + rectangle B
      + rectangle C + $\frac{1}{4}$ circle D

Shape (iv) = triangle A + rectangle B + semi-circle C

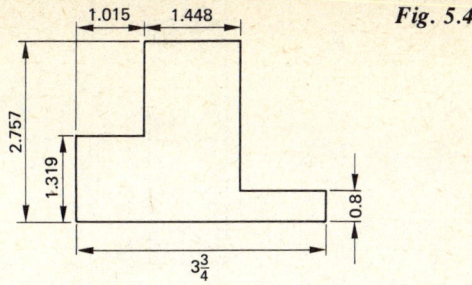

Fig. 5.43

Exercises 5

5.1 Convert:
a) 3.78 km to m   *b*) 1.043 m to mm
c) 89 mm to cm   *d*) 2 640 mm to m
e) 1.3 m to cm   *f*) 4 965 m to km
g) 0.95 m to mm   *h*) 675 cm to m

5.2 Convert:
a) 3.7 inches to mm   *b*) 12 yards to m
c) 0.91 inch to mm   *d*) 6.45 inches to mm
e) 15 miles to km   *f*) $3\frac{1}{2}$ inches to mm
g) 14.5 inches to cm   *h*) $1\frac{1}{2}$ feet to m

5.3 Convert to inches:
a) 38.1 mm *b*) 91.44 mm *c*) 13.97 mm *d*) 317.5 mm
e) 125.73 mm *f*) 6.35 cm *g*) 355.6 mm *h*) 7.62 mm

5.4 Convert to millimetres; give the answer correct to 2 decimal places:
a) 2.72 inches *b*) 8.63 inches *c*) 1.755 inches
d) $4\frac{3}{4}$ inches *e*) 0.817 inch *f*) 3.076 inches
g) 0.005 inch *h*) 15.658 inches

5.5 The dimensions of the shaft shown in fig 5.42 are given in inches. Sketch the shaft and give its dimensions in millimetres correct to the nearest 0.1 mm.

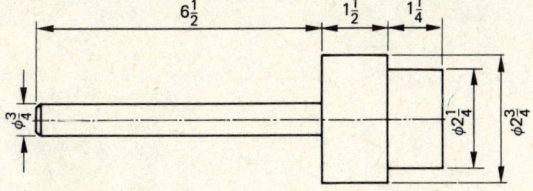

Fig. 5.42

5.6 The dimensions of the machined block shown in fig. 5.43 are given in inches. Sketch the block and give its dimensions in millimetres correct to the nearest 0.01 mm.

5.7 Find the width in millimetres of the tenon shown in fig. 5.44 if it is to fit the slot with 0.02 mm clearance.

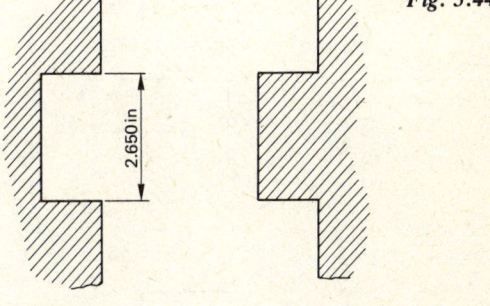

Fig. 5.44

5.8 What diameter of shaft in millimetres would run in a bearing 3.75 inches diameter and give a clearance of 0.05 mm?

5.9 In each of the pairs of mating components shown in fig. 5.45, component A is required to fit with component B with the stated clearance between the mating parts. Give the missing dimensions for each component B in millimetres correct to the nearest 0.01 mm.

5.10 A boring bar is designed to use $\frac{3}{8}$ inch square tool bits. The nearest size in metric tool bits available is 10 mm square. Calculate the increase in cross-sectional area using the metric tool bits. Give the answer correct to the nearest 0.1 mm².

5.11 Convert:
a) 1 450 000 m² to km²   *b*) 73 000 cm² to m²
c) 2 918 000 mm² to m²   *d*) 4.5 km² to m²
e) 1 000 mm² to cm²   *f*) 147 cm² to mm²

5.12 Giving the answers correct to 2 decimal places, convert:
a) 6.3 square inches to mm²
b) 1 400 mm² to square inches
c) 8.5 square inches to cm²

5.13 Determine the area and perimeter of each of the following rectangles:
a) 30 mm by 10 mm   *b*) 6.5 cm by 4.5 cm
c) 340 mm by 180 mm   *d*) 1.75 m by 0.8 m
e) 17.5 mm by 13.4 mm   *f*) 79 mm by 56 mm
g) 419 mm by 30.5 cm   *h*) 1.05 mm by 2.6 mm

Length and Area 51

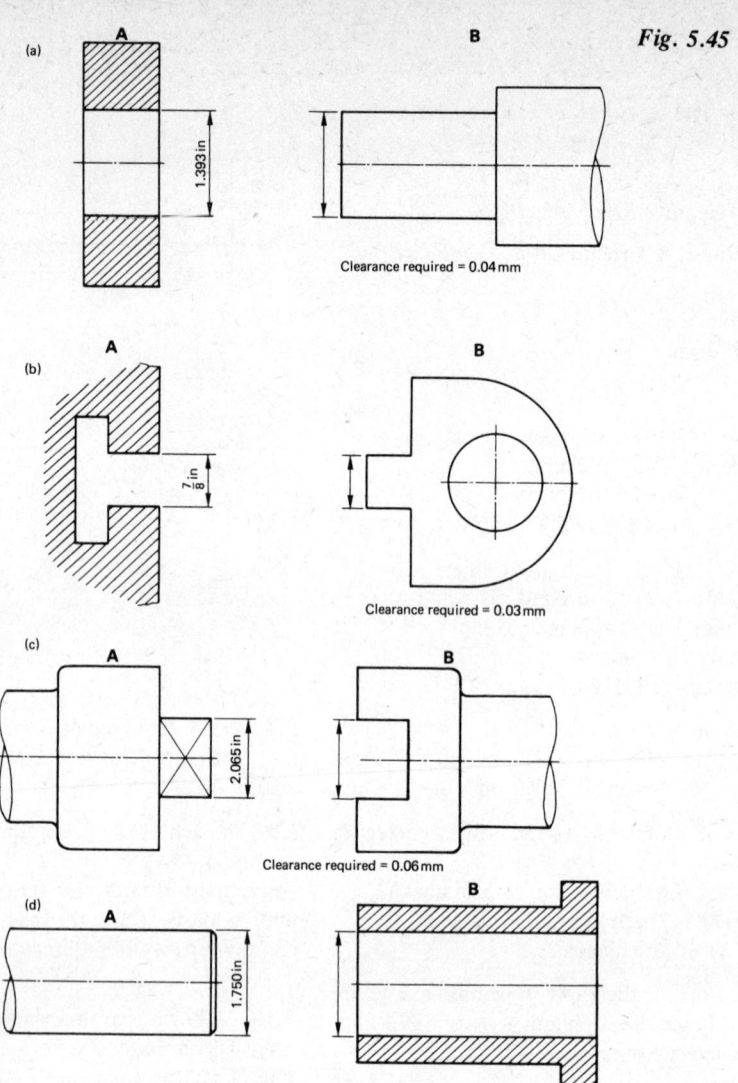

Fig. 5.45

(a)

1.393 in

B

Clearance required = 0.04 mm

(b)

A

7/8 in

B

Clearance required = 0.03 mm

(c)

A

2.065 in

B

Clearance required = 0.06 mm

(d)

A

1.750 in

B

Clearance required = 0.02 mm

5.14 Calculate the area of each of the following:
a) a metal plate 315 mm by 75 mm
b) a square surface plate of 250 mm side
c) the cross-section of 20 mm square bar
d) the floor area of a rectangular workshop 12 m by 10.5 m
e) the cross-section of a timber plank 25 cm by 3.8 cm
f) a rectangular machine base 2.54 m by 1.29 m
g) a bench top 2 m by 600 mm (answer in m²)
h) a scribing block base 65 mm by 55 mm

5.15 Fig. 5.46 shows the floor plan of a service garage.
a) Find the area and perimeter of each of the following: (i) Service bay, (ii) Office, (iii) Reception, (iv) Parts store, (v) Workshop.
b) What is the total floor area of the garage?

5.16 Fig. 5.47 shows a ventilation panel having three rectangular slots. Give the area of the panel in square millimetres.

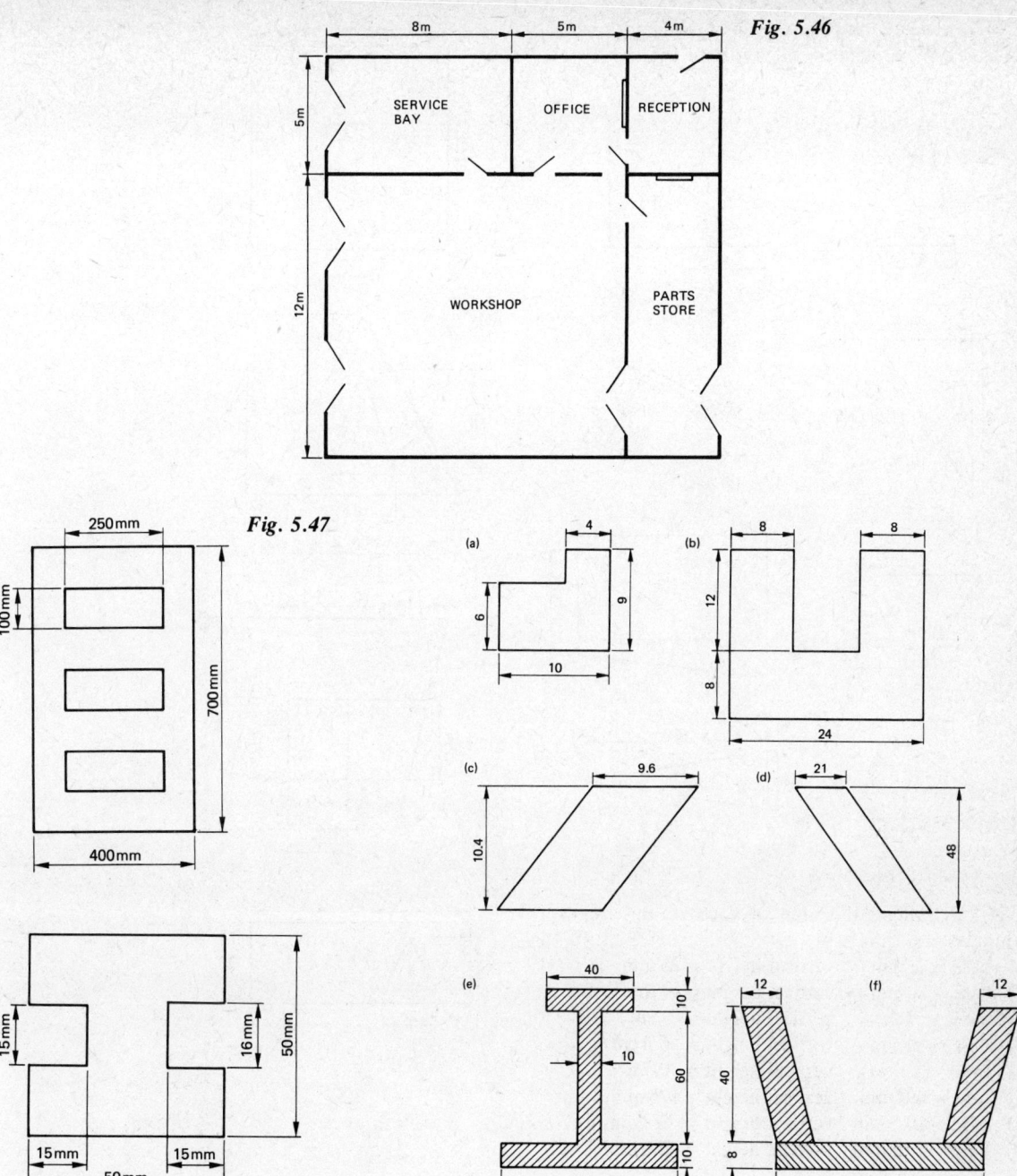

Fig. 5.46

Fig. 5.47

Fig. 5.48

Fig. 5.49

5.17 Calculate the area of the plate gauge shown in fig. 5.48.

5.18 Calculate the area of each of the shapes shown in fig. 5.49. All dimensions are in millimetres.

5.19 In a sliding operation using a knife tool on a centre lathe, the longitudinal feed is 0.35 mm/rev and the depth of cut is 5.6 mm. Calculate the area of the cut.

5.20 Calculate the area of each of the triangles shown in fig. 5.50. All dimensions are in millimetres.

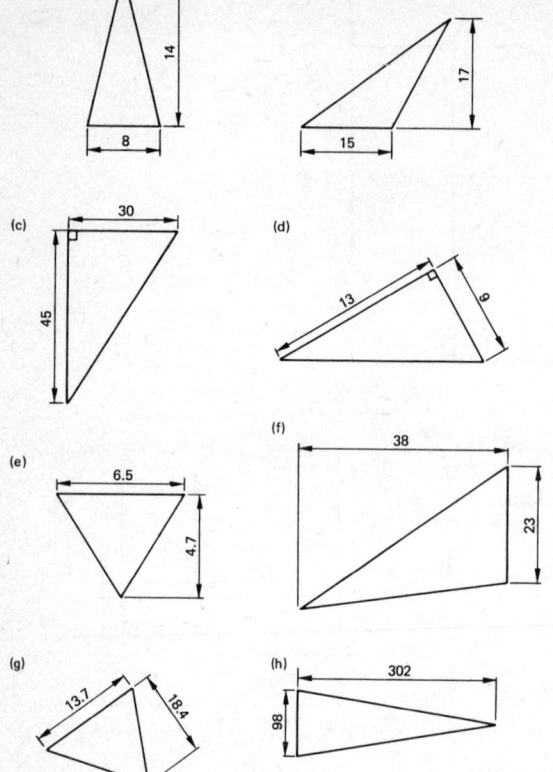

Fig. 5.50

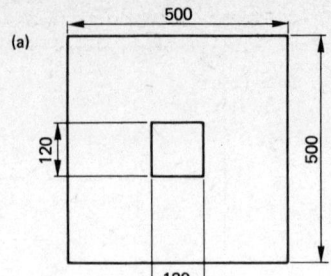

Fig. 5.51

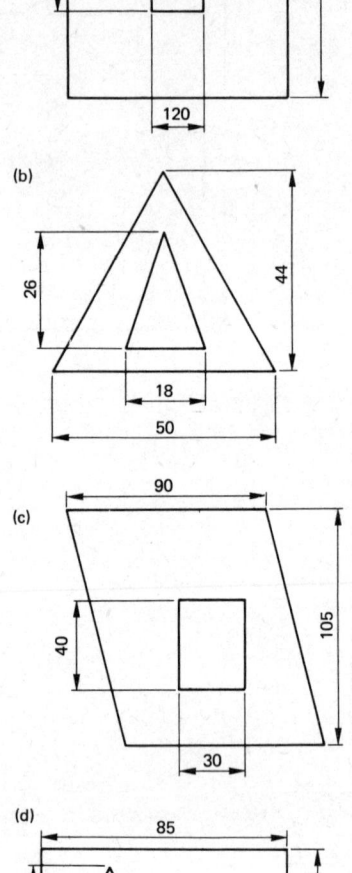

5.21 Determine the area of each of the following triangles:

a) base = 11 mm   vertical height = 23 mm
b) base = 80 mm   vertical height = 44 mm
c) base = 9 cm   vertical height = 7 cm
d) base = 80 mm   vertical height = 6.4 cm
e) base = 2.1 m   vertical height = 0.7 m
f) base = 494 mm   vertical height = 176 mm
g) base = 10.5 mm vertical height = 9.3 mm
h) base = 0.8 mm vertical height = 2.6 mm

5.22 Each of the shapes in fig. 5.51 shows a plate of metal which has a hole punched out. Determine the remaining area of the plate in each case. All dimensions are in millimetres.

5.23 The circumference of the following circles is required; take $\pi = \frac{22}{7}$.

a) 84 mm diameter *b*) 98 mm radius
c) 147 mm diameter *d*) 42 cm diameter
e) 175 cm radius. *f*) 3.5 m diameter
g) 10.5 mm diameter *h*) 157.5 mm diameter

5.24 Calculate the circumference of the following circles; take $\pi = 3.142$, and give the answer correct to 2 decimal places.

a) 33.2 mm diameter *b*) 187.4 mm diameter
c) 3.19 cm diameter *d*) 14.6 mm diameter
e) 25.4 mm diameter *f*) 16.3 mm radius
g) 1.09 m radius *h*) 49.3 mm diameter

5.25 The area of the following circles is required; take $\pi = \frac{22}{7}$.
a) 7 m radius *b*) 21 mm diameter
c) 1.4 mm diameter *d*) 35 mm diameter
e) 28 cm diameter *f*) 147 mm radius
g) 700 mm diameter *h*) 154 mm diameter

5.26 Calculate the area of the following circles. Take $\pi = 3.142$, and give the answer correct to 4 significant figures.
a) 38.5 mm diameter *b*) 464 mm diameter
c) 124.8 mm diameter *d*) 59.6 cm diameter
e) 1.46 m diameter *f*) 47.25 mm radius
g) 83 mm diameter *h*) 519 mm diameter

5.27 Determine the cross-sectional area of a steel bush 41 mm outside diameter and 29 mm inside diameter. Take $\pi = 3.142$, and give the answer correct to one decimal place.

5.28 Calculate the area of the cover plate shown in fig. 5.52. Take $\pi = \frac{22}{7}$.

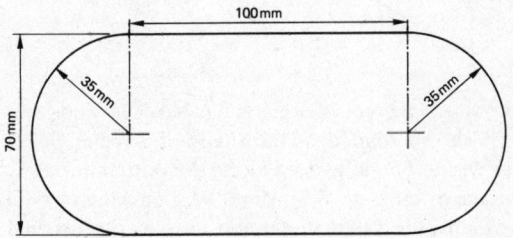

Fig. 5.52

5.29 Find the area of the drilled plate shown in fig. 5.53.

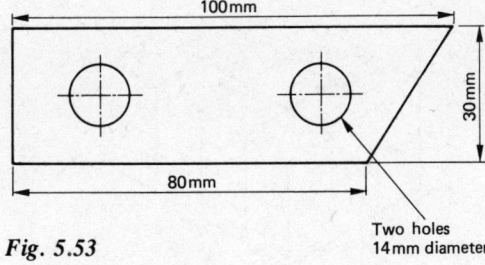

Fig. 5.53

5.30 Calculate the area of the end face of the vee-block shown in fig. 5.54

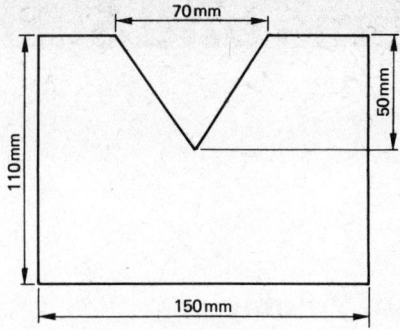

Fig. 5.54

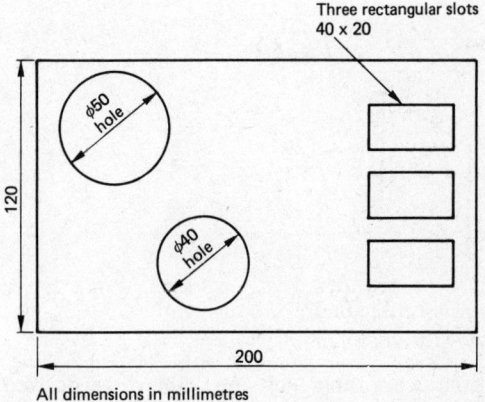

All dimensions in millimetres

Fig. 5.55

5.31 Complete the following table to show metric diameters of copper fuse wire:

FUSING CURRENT amps	DIAMETER inches	DIAMETER mm
3	0.0044	
5	0.0062	
10	0.0098	
15	0.0129	
30	0.0200	

5.32 Fig. 5.55 shows the baseplate of an instrument chassis. Calculate its area in square millimetres.

6 Volume and Surface Area

6.1 Units of Volume

The **volume of a solid** is given by the number of **cubic units** that it contains.

1 cubic metre is defined as the volume contained by a cube of 1 m side.

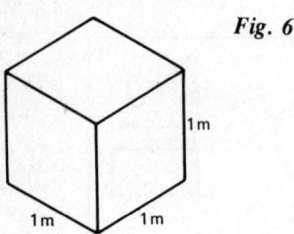

Fig. 6.1

Volume = 1 cubic metre
usually written as 1 m^3

In the same way, other units of volume are derived from the units of length.

Volume = 1 cubic kilometre = 1 km^3

Volume = 1 cubic millimetre = 1 mm^3

Volume = 1 cubic centimetre = 1 cm^3

Fig. 6.2

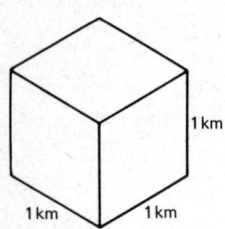

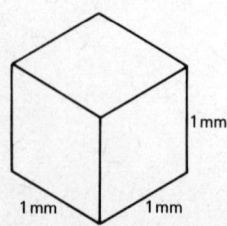

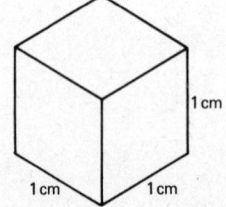

6.2 Volume of a Prism

A **prism** is a solid figure which has constant shape and constant area of cross-section throughout its length.

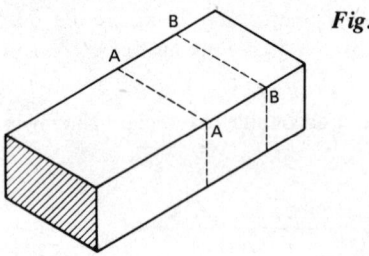

Fig. 6.3

Fig. 6.3 shows a piece of rectangular bar. The end face, which is shown shaded, is the shape of a rectangle. If the bar was to be guillotined along the dotted line AA, a new face or cross-section would be seen which would have exactly the same shape and area as the original end face. This would also apply for a cut at BB or at any other point along the length of the bar.

Thus the bar has a constant cross-section and may be described as a rectangular prism.

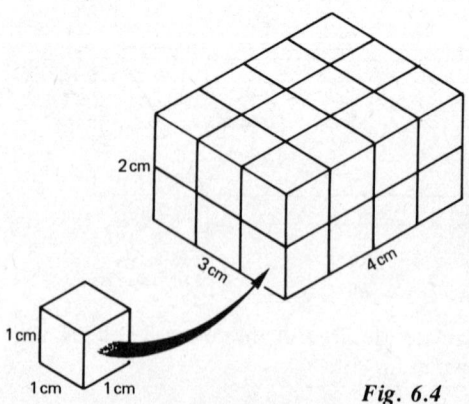

Fig. 6.4

Fig. 6.4 shows a rectangular prism which is seen to be made up of two layers, each of 12 cubes of 1 cm side.

Hence, the prism contains 24 cubes of 1 cm side, i.e.

Volume of the prism = 24 cubic centimetres
= 24 cm^3

Area of the end face $= 2\,\text{cm} \times 3\,\text{cm}$
$$= 6\,\text{cm}^2$$
Volume $=$ area of end face $\times$ length
$$= 6\,\text{cm}^2 \times 4\,\text{cm}$$
$$= 24\,\text{cm}^3$$

This general formula can be used to find the volume of any shape of prism when the dimensions are in the same units:

Volume of prism $=$ area of end face $\times$ length

Example 6.1 Calculate the volume of the rectangular prism shown in fig. 6.5.

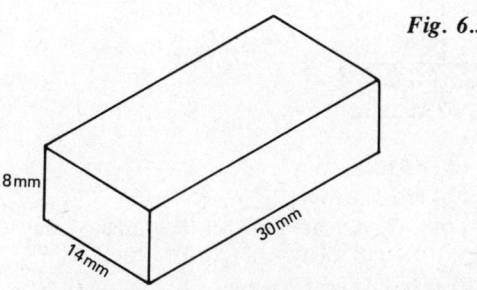

Fig. 6.5

Area of end face $= 14\,\text{mm} \times 8\,\text{mm} = 112\,\text{mm}^2$
Volume of prism $=$ area of end face $\times$ length
$$= 112\,\text{mm}^2 \times 30\,\text{mm}$$
$$= 3\,360\,\text{mm}^3 \quad (Ans.)$$

Example 6.2 Calculate the volume of the rectangular block of metal shown in fig. 6.6.

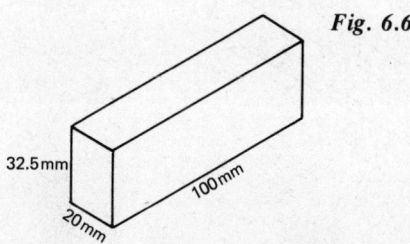

Fig. 6.6

Area of end face $= 20 \times 32.5 = 650\,\text{mm}^2$
Volume $= 650\,\text{mm}^2 \times 100\,\text{mm} = 65\,000\,\text{mm}^3 \quad (Ans.)$

Example 6.3 Calculate the volume of a steel block of length 150 mm, width 30 mm and height 20 mm.

Sketching the block (fig. 6.7)

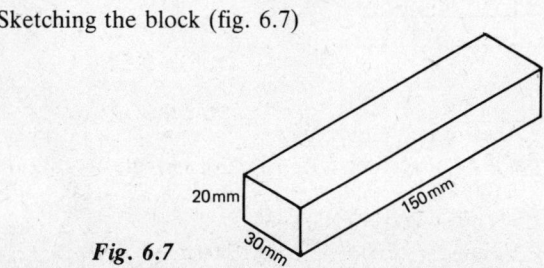

Fig. 6.7

Area of end face $= 30 \times 20 = 600\,\text{mm}^2$
Volume $= 600\,\text{mm}^2 \times 150\,\text{mm} = 90\,000\,\text{mm}^3 \quad (Ans.)$

Example 6.4 Find the volume of a block of metal $40\,\text{mm} \times 50\,\text{mm} \times 80\,\text{mm}$.
The question does not state which dimension is the length or width or height. This is of no consequence as the block will have the same volume regardless of which face is considered to be the end face. Fig. 6.8 shows two possible positions in which the block may be drawn.

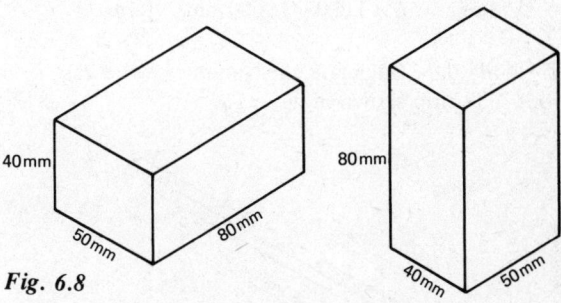

Fig. 6.8

Area of end face $= 50 \times 40 = 2\,000\,\text{mm}^2$
Volume $= 2\,000 \times 80 = 160\,000\,\text{mm}^3$

Area of end face $= 40 \times 80 = 3\,200\,\text{mm}^2$
Volume $= 3\,200 \times 50 = 160\,000\,\text{mm}^3$

Example 6.5 Calculate the volume of a 20 cm length of 10 mm square bar.
Area of end face $= 10 \times 10 = 100\,\text{mm}^2$
Length $= 20\,\text{cm} = 200\,\text{mm}$
Volume $= 100 \times 200 = 20\,000\,\text{mm}^3 \quad (Ans.)$

Example 6.6 Calculate the volume of the triangular prism shown in fig. 6.9.

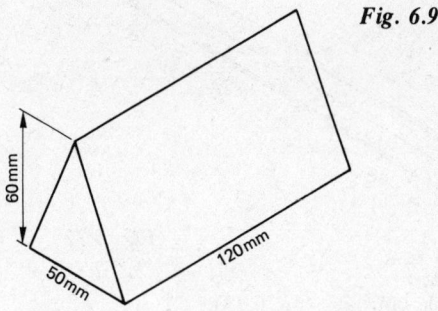

Fig. 6.9

Area of end face $=$ area of triangle
$$= \tfrac{1}{2} \times 50 \times 60 = 1\,500\,\text{mm}^2$$
Volume $= 1\,500 \times 120 = 180\,000\,\text{mm}^3 \quad (Ans.)$

Example 6.7 Calculate the volume of a 1 m length of aluminium extrusion having the cross-section shown in fig. 6.10.

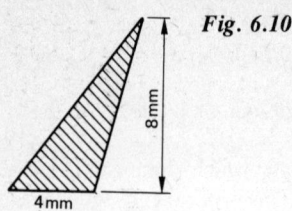

Fig. 6.10

Area of end face = area of cross-section
$$= \tfrac{1}{2} \times 4 \times 8 = 16\,\text{mm}^2$$
Length = 1 m = 1 000 mm
Volume = 16 × 1 000 = 16 000 mm³   (*Ans.*)

Example 6.8   Calculate the volume of the cast iron tool-slide strip shown in fig. 6.11.

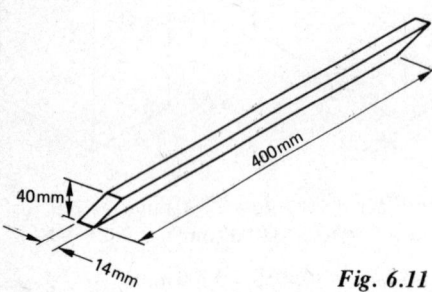

Fig. 6.11

Area of end face = area of parallelogram
$$= 14 \times 40 = 560\,\text{mm}^2$$
Volume = 560 × 400 = 224 000 mm³   (*Ans.*)

Example 6.9   Calculate the volume of the length of angle iron shown in fig. 6.12.

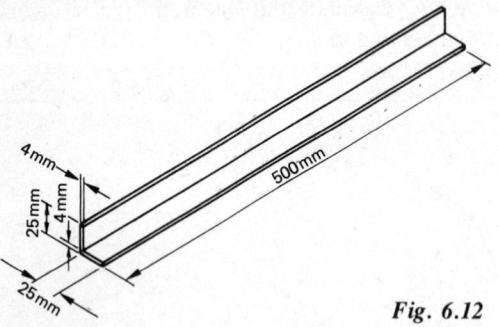

Fig. 6.12

Sketching the end face (fig. 6.13),

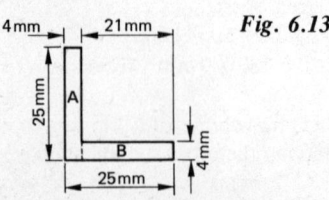

Fig. 6.13

Area of end face = area of rect. A + area of rect. B
$$= (4 \times 25) + (21 \times 4)$$
$$= 100 + 84 = 184\,\text{mm}^2$$
Volume = 184 × 500 = 92 000 mm³   (*Ans.*)

Example 6.10   A fabricated beam has the cross-section shown in fig. 6.14. Calculate the volume of a 2 m length of the beam.

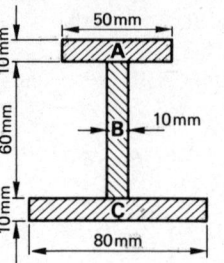

Fig. 6.14

Area of end face
= area of cross-section
= area of rect. A + area of rect. B + area of rect. C
$$= (50 \times 10) + (10 \times 60) + (80 \times 10)$$
$$= 500 + 600 + 800 = 1\,900\,\text{mm}^2$$
Length = 2 m = 2 000 mm
Volume = 1 900 × 2 000 = 3 800 000 mm³   (*Ans.*)

Example 6.11   Calculate the volume of the cast-iron adjusting strip shown in fig. 6.15.

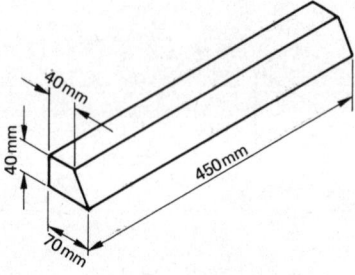

Fig. 6.15

Sketching the end face (fig. 6.16),

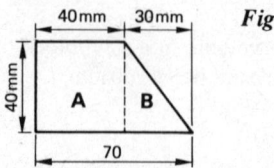

Fig. 6.16

Area of end face
= area of rectangle A + area of triangle B
$$= (40 \times 40) + (\tfrac{1}{2} \times 30 \times 40)$$
$$= 1\,600 + 600 = 2\,200\,\text{mm}^2$$
Volume = 2 200 × 450 = 990 000 mm³   (*Ans.*)

6.3 Volume of a Cylinder

A **cylinder** is a prism whose constant cross-section is a circle.

Volume of a cylinder = area of end face × length where area of end face = πr^2.

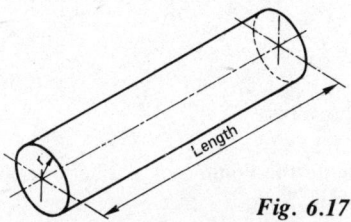

Fig. 6.17

Example 6.12 Calculate the volume of a cylinder having a radius of 14 mm and a length of 200 mm. Take $\pi = \frac{22}{7}$.

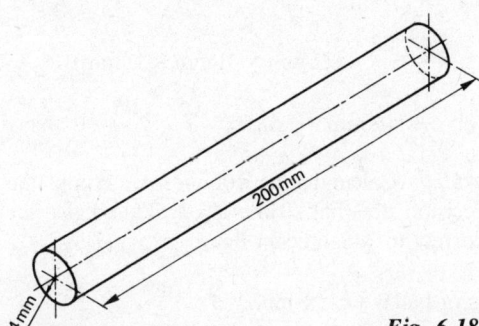

Fig. 6.18

Area of end face = πr^2

$$= \frac{22}{7} \times 14 \times 14 = 616 \, \text{mm}^2$$

Volume = area of face × length
= 616 × 200 = 123 200 mm³   (*Ans.*)

Example 6.13 Calculate the volume of a cylinder having a radius of 4 cm and a length of 10.7 cm. Take $\pi = 3.142$.

Area of end face = πr^2
= 3.142 × 4 × 4 = 50.272 cm²

Volume = area of end face × length
= 50.272 × 10.7 = 537.91 cm³   (*Ans.*)

Example 6.14 Calculate the volume of a cylinder 30 mm diameter and 100 mm long. Take $\pi = 3.142$.

Radius of cylinder = $\frac{30}{2}$ = 15 mm

Area of end face = πr^2
= 3.142 × 15 × 15 = 706.95 mm²

Volume = 706.95 × 100 = 70 695 mm³   (*Ans.*)

Example 6.15 Calculate the volume in cubic millimetres of a 300 mm length of 18 mm diameter bar. Take $\pi = 3.142$ and give the answer correct to 3 significant figures.

Radius of bar = $\frac{18}{2}$ = 9 mm

Area of end face = 3.142 × 9 × 9 = 254.5 mm²
Volume = 254.5 × 300
= 76 350.6 = 76 400 mm³ to 3 s.f.   (*Ans.*)

Example 6.16 Calculate the volume of the shaft shown in fig. 6.19. Take $\pi = 3.142$ and give the answer correct to 4 significant figures.

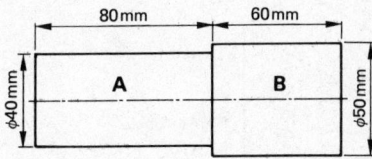

Fig. 6.19

Vol. of shaft = vol. of cylinder A + vol. of cylinder B

Radius of cylinder A = $\frac{40}{2}$ = 20 mm

Area of end face of cylinder A
= 3.142 × 20 × 20 = 1 256.8 mm²
Vol. of cylinder A = 1 256.8 × 80 = 100 544 mm³

Radius of cylinder B = $\frac{50}{2}$ = 25 mm

Area of end face of cylinder B
= 3.142 × 25 × 25 = 1 963.75 mm²
Vol. of cylinder B = 1 963.75 × 60 = 117 825 mm²

Vol. of shaft = 100 544 + 117 825
= 218 369 mm³ = 218 400 mm³ to 4 s.f.
(*Ans.*)

Example 6.17 Calculate the volume of the cast iron bush shown in fig. 6.20. Take $\pi = 3.142$ and give the answer correct to 3 significant figures.

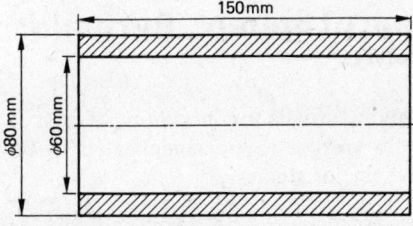

Fig. 6.20

Radius of outer circle $= \dfrac{80}{2} = 40\,\text{mm}$

Radius of inner circle $= \dfrac{60}{2} = 30\,\text{mm}$

  Area of end face
  = area of outer circle − area of inner circle
  $= (3.142 \times 40 \times 40) - (3.142 \times 30 \times 30)$
  $= 5\,027.2 - 2\,827.8 = 2\,199.4\,\text{mm}^2$

  Volume $= 2\,199.4 \times 150$
          $= 329\,910\,\text{mm}^3 = 330\,000\,\text{mm}^3$ to 3 s.f. *(Ans.)*

Example 6.18 Calculate the volume of the drilled plate shown in fig. 6.21. Take $\pi = \frac{22}{7}$.

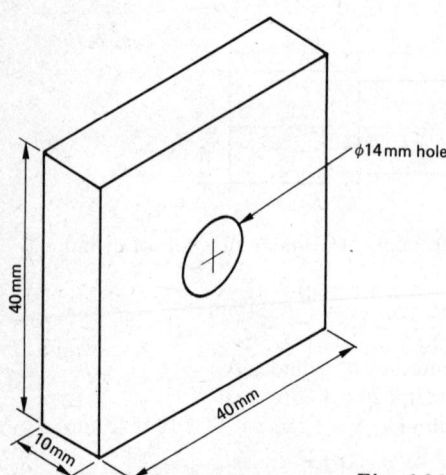

Fig. 6.21

Consider the 40 mm square face as the end face so that the hole is included. Therefore

  Area of end face
  = area of plate − area of hole

$$= (40 \times 40) - \left(\dfrac{22}{7} \times 7 \times 7\right)$$

  $= 1\,600 - 154 = 1\,446\,\text{mm}^2$
  Volume $= 1\,446 \times 10 = 14\,460\,\text{mm}^3$ *(Ans.)*

6.4 Volume of Sphere, Pyramid and Cone

The derivation of the formula for the volume of each of the shapes in this section is too complicated to be discussed at this stage of studies.

1 A **sphere** is a solid figure having the shape of a ball, e.g. a ball bearing is a sphere.

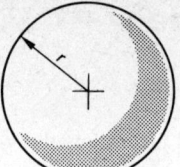

Fig. 6.22

Volume of a sphere $= \dfrac{4}{3}\pi r^3$

where r = radius of the sphere

Example 6.19 Calculate the volume of a sphere having a diameter of 42 mm. Take $\pi = \frac{22}{7}$.

  Radius of sphere $= \dfrac{42}{2} = 21\,\text{mm}$

  Volume $= \dfrac{4}{3}\pi r^3$

          $= \dfrac{4}{3} \times \dfrac{22}{7} \times (21\,\text{mm} \times 21\,\text{mm} \times 21\,\text{mm})$

          $= 38\,808\,\text{mm}^3$ *(Ans.)*

Example 6.20 Calculate the volume of a 5 mm diameter precision steel ball. Take $\pi = 3.142$ and give the answer correct to 3 significant figures.

  Radius of ball $= \dfrac{5}{2} = 2.5\,\text{mm}$

  Volume $= \dfrac{4}{3} \times 3.142 \times (2.5 \times 2.5 \times 2.5)$

          $= 65.46\,\text{mm}^3 = 65.5\,\text{mm}^3$ to 3 s.f.   *(Ans.)*

2 Fig. 6.23 shows a **square pyramid**. The type of a pyramid is defined by the shape of its base, so that a square pyramid is one having a square base.

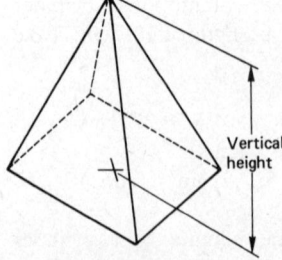

Fig. 6.23

The volume of a pyramid is given by

  Volume of pyramid

  $= \dfrac{1}{3}$ area of the base × vertical height

Example 6.21 Calculate the volume of a pyramid which has a base 20 mm square and a vertical height of 30 mm.

Area of base $= 20 \times 20 = 400 \, \text{mm}^2$

$$\text{Volume} = \frac{1}{3} \text{ area of base } \times \text{vertical height}$$

$$= \frac{1}{3} \times 400 \, \text{mm}^2 \times 30 \, \text{mm} = 4\,000 \, \text{mm}^3 \quad (Ans.)$$

Example 6.22 Calculate the volume of a pyramid which has a rectangular base 4 cm by 3 cm and a vertical height of 6 cm.

Area of base $= 4 \times 3 = 12 \, \text{cm}^2$

$$\text{Volume} = \frac{1}{3} \text{ area of base } \times \text{vertical height}$$

$$= \frac{1}{3} \times 12 \, \text{cm}^2 \times 6 \, \text{cm} = 24 \, \text{cm}^3 \quad (Ans.)$$

3 The **cone** shown in fig. 6.24 may be considered as a pyramid with a circular base.

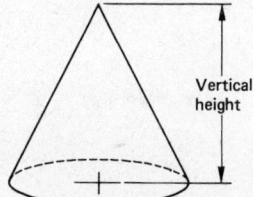

Fig. 6.24

Vertical height

$$\text{Volume of cone} = \frac{1}{3} \text{ area of base } \times \text{vertical height}$$

where the area of the base is the area of a circle of radius r.

Example 6.23 Calculate the volume of a cone having a base of 14 cm diameter and a vertical height of 24 cm. Take $\pi = \frac{22}{7}$.

$$\text{Radius of base} = \frac{14}{2} = 7 \, \text{cm}$$

$$\text{Area of base} = \pi r^2 = \frac{22}{7} \times 7 \times 7 = 154 \, \text{cm}^2$$

$$\text{Volume of cone} = \frac{1}{3} \text{ area of base } \times \text{vertical height}$$

$$= \frac{1}{3} \times 154 \, \text{cm}^2 \times 24 \, \text{cm}$$

$$= 1\,232 \, \text{cm}^3 \quad (Ans.)$$

Example 6.24 Calculate the volume of a cone of base 40 mm diameter and vertical height 50 mm. Take $\pi = 3.142$ and give the volume correct to 4 significant figures.

$$\text{Radius of base} = \frac{40}{2} = 20 \, \text{mm}$$

Area of base $= \pi r^2 = 3.142 \times 20 \times 20 = 1\,256.8 \, \text{mm}^2$

$$\text{Volume of cone} = \frac{1}{3} \text{ area of base } \times \text{vertical height}$$

$$= \frac{1}{3} \times 1\,256.8 \times 50$$

$$= 20\,947 \, \text{mm}^3$$
$$= 20\,950 \, \text{mm}^3 \text{ to 4 s.f.} \quad (Ans.)$$

6.5 Volume of Containers

1 It is often required to find the volume which can be held by a container such as a box, crate or storage vessel. This may be done by considering an **imaginary solid** of the same size as the internal dimensions of the container. The volume which can be held by the container is then equal to the volume of the imaginary solid.

Example 6.25 Determine the volume in cubic metres that could be contained by a rectangular crate of internal dimensions:

length 2 m   width 1.5 m   height 1.8 m

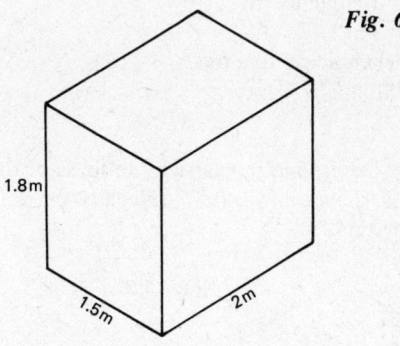

Fig. 6.25

1.8 m

1.5 m

2 m

Fig. 6.25 shows the imaginary solid of the same size as the internal dimensions of the crate.

Area of end face $= 1.5 \times 1.8 = 2.7 \, \text{m}^2$
Volume $= 2.7 \times 2 = 5.4 \, \text{m}^3$

Therefore, the volume that could be contained by the crate $= 5.4 \, \text{m}^3 \quad (Ans.)$

2 When the thickness of the walls of the container is small in relation to its other dimensions, e.g. a sheet metal box, then the difference between **the internal and the external dimensions** may be neglected when finding the volume that can be contained.

Example 6.26 Calculate the volume in cubic millimetres that could be contained by a sheet metal box 200 mm long, 80 mm wide and 70 mm high. (Neglect the thickness of the sheet metal.)
Sketching the imaginary solid (fig. 6.26)

Fig. 6.26

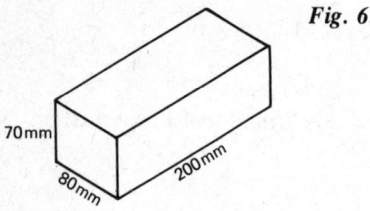

Area of end face = $80 \times 70 = 5\,600$ mm²
Volume = $5\,600 \times 200 = 1\,120\,000$ mm³   (*Ans.*)

Example 6.27 Calculate the volume to the nearest cubic centimetre that could be contained in a metal tool box 35 cm long, 22.5 cm wide and 27.5 cm high. (Neglect the thickness of the metal.)

Area of end face = $22.5 \times 27.5 = 618.75$ cm²
Volume = $618.75 \times 35 = 21\,656$ cm³   (*Ans.*)

3 When a container is intended to hold liquids, the total volume it could contain is referred to as its **capacity** and may be given in litres.

1 000 litres = 1 cubic metre
1000 l = 1 m³
1 litre = 1 000 cubic centimetres
1 l = 1 000 cm³

Example 6.28 Determine the capacity in litres of the oil storage tank shown in fig. 6.27. (Neglect the wall thickness of the tank.)

Fig. 6.27

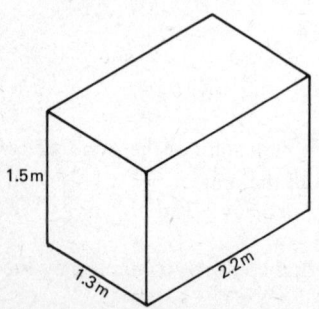

Area of end face = $1.3 \times 1.5 = 1.95$ m²
Volume = $1.95 \times 2.2 = 4.29$ m³

1 m³ = 1 000 l
∴ Capacity of tank = $4.29 \times 1\,000$
$= 4\,290$ litres   (*Ans.*)

Example 6.29 How many litres of liquid fertiliser could be contained in a drum of internal dimensions 0.6 m diameter and 0.8 m height? Take $\pi = 3.142$ and give the capacity to the nearest litre.

$$\text{Radius of drum} = \frac{0.6}{2} = 0.3 \text{ m}$$

Area of end face = πr^2
$= 3.142 \times 0.3 \times 0.3 = 0.2828$ m²
Volume = $0.2828 \times 0.8 = 0.2262$ m³

Capacity = $0.2262 \times 1\,000 = 226.21$
Drum could contain 226 litres   (*Ans.*)

Example 6.30 Determine how much water could be contained in the windscreen washer reservoir shown in fig. 6.28. Neglect the wall thickness and give the capacity to the nearest 0.1 l.

Fig. 6.28

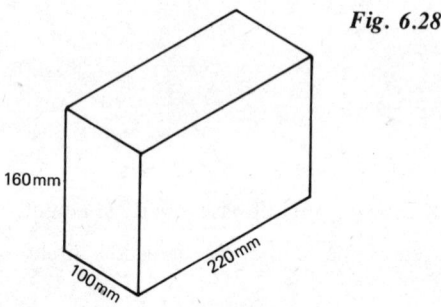

For convenience change the dimensions to centimetres:
Width = 100 mm = 10 cm
Height = 160 mm = 16 cm
Length = 220 mm = 22 cm

Area of end face = $10 \times 16 = 160$ cm²
Volume = $160 \times 22 = 3\,520$ cm³

1 litre = 1 000 cm³

∴ Capacity $= \dfrac{3\,520}{1\,000} = 3.521$

Reservoir holds 3.5 litres   (*Ans.*)

6.6 Volume of Metal Removed in Machining

The volume of metal removed by a machining operation may be found by either of the following methods:
1. Comparing the volumes of the component before and after the machining process.
2. Calculating the volume of the shape removed by machining.

Consider a piece of 28 mm diameter bar 100 mm long which is turned down to 21 mm diameter throughout its length.

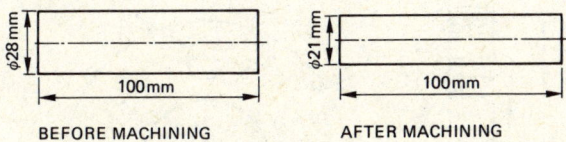

BEFORE MACHINING   AFTER MACHINING

Fig. 6.29

Method 1
The volume before machining
= area of end face × length

Area of end face $= \dfrac{22}{7} \times 14 \times 14 = 616 \text{ mm}^2$

∴ Volume before machining $= 616 \times 100$
$= 61\,600 \text{ mm}^3$

The volume after machining
= area of end face × length

Area of end face $= \dfrac{22}{7} \times 10.5 \times 10.5 = 346.5 \text{ mm}^2$

∴ Volume after machining $= 346.5 \times 100$
$= 34\,650 \text{ mm}^3$

Volume of metal removed
= vol. before machining − vol. after machining
= 61 600 − 34 650 = 26 950 mm³ (*Ans.*)

Method 2 The shaded portion in fig. 6.30 represents the tube of metal that has been machined away to

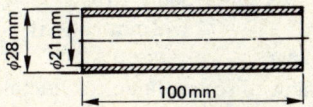

Fig. 6.30

produce the 21 mm diameter. Therefore the volume of the shaded portion is the volume of metal removed by machining.

Area of end face $= \left(\dfrac{22}{7} \times 14 \times 14\right) - \left(\dfrac{22}{7} \times 10.5 \times 10.5\right)$

$= 616 - 346.5$
$= 269.5 \text{ mm}^2$

Volume of metal removed $= 269.5 \times 100$
$= 26\,950 \text{ mm}^3$ (*Ans.*)

Example 6.31 Fig. 6.31 shows a cast iron block before and after machining. Calculate the volume of metal removed by machining.

Fig. 6.31

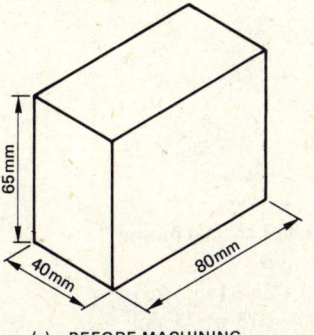

(a) BEFORE MACHINING

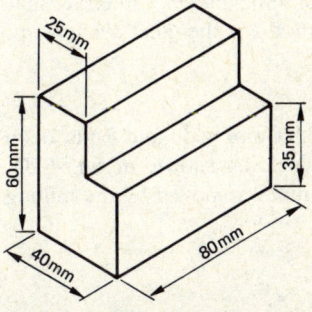

(b) AFTER MACHINING

Method 1
Volume before machining
= area of end face × length
Area of end face = 40 × 65 = 2 600 mm²
Volume before machining
= 2 600 × 80 = 208 000 mm³
Volume after machining
= area of end face × length
Area of end face = (25 × 60) + (15 × 35)
$= 1\,500 + 525 = 2\,025 \text{ mm}^2$
Volume after machining = 2 025 × 80 = 162 000 mm³

Volume of metal removed
= vol. before machining − vol. after machining
= 208 000 − 162 000 = 46 000 mm³ (*Ans.*)

Method 2 Fig. 6.32 shows the shape of the metal which is removed by the machining operation.

Volume and Surface Area 63

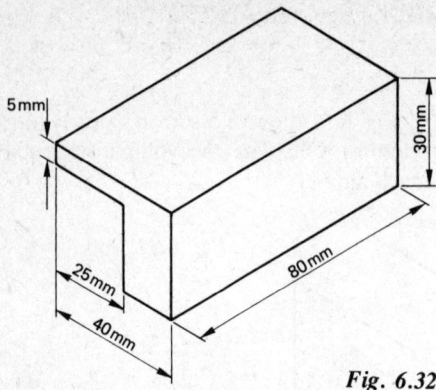

5mm

30mm

80mm

25mm

40mm

Fig. 6.32

Volume of metal removed by machining
= area of end face × length
Area of end face = (40 × 5) + (15 × 25)
$$= 200 + 375 = 575 \text{ mm}^2$$
Volume of metal removed = 575 × 80
$$= 46\,000 \text{ mm}^3 \quad (Ans.)$$
The two methods give the same answer and either may be used, but usually Method 2 is the most convenient.

Example 6.32 Two slots 10 mm wide and 8 mm deep are milled in a steel block as shown in fig. 6.33. Calculate the volume of metal removed by the milling operation.

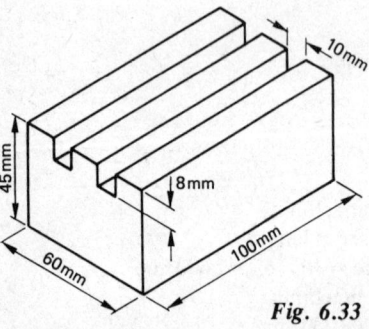

10mm

45mm

8mm

60mm

100mm

Fig. 6.33

Using Method 2, and sketching the shape of the metal removed (fig. 6.34),

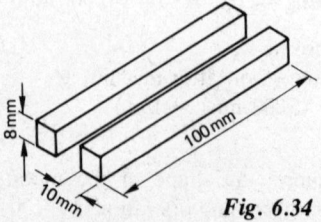

8mm

100mm

10mm

Fig. 6.34

Area of end face (1 slot) = 10 × 8 = 80 mm²
Volume removed (1 slot) = 80 × 100
$$= 8\,000 \text{ mm}^3$$
Volume removed (2 slots) = 2 × 8 000
$$= 16\,000 \text{ mm}^3 \quad (Ans.)$$

Example 6.33 *a*) Calculate the volume of metal removed in boring the 60 mm hole and drilling the four 16 mm holes in the steel flange shown in fig. 6.35.
b) Give the volume of the finished flange to the nearest 100 cubic millimetres.

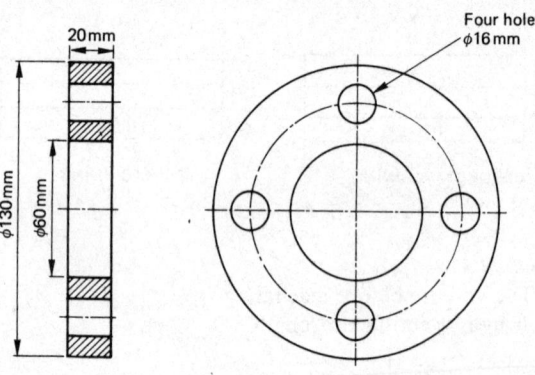

20mm

ϕ130mm

ϕ60mm

Four holes
ϕ16 mm

Fig. 6.35

a) Volume of metal removed
= volume of 60 mm bore + volume of four 16 mm holes
Area of 60 mm bore = πr^2
$$= 3.142 × 30 × 30$$
$$= 2\,828 \text{ mm}^2$$
Volume of 60 mm bore = 2 828 × 20
$$= 56\,560 \text{ mm}^3$$
Area of one 16 mm hole = πr^2
$$= 3.142 × 8 × 8$$
$$= 201.1 \text{ mm}^2$$
Volume of one 16 mm hole = 201.1 × 20
$$= 4\,022 \text{ mm}^3$$
Volume of four 16 mm holes = 4 022 × 4
$$= 16\,088 \text{ mm}^3$$
∴ Volume of metal removed = 56 560 + 16 088
$$= 72\,648 \text{ mm}^3 \quad (Ans.)$$
b) Volume of finished flange
= volume of the uncut disc − volume of metal removed
Area of uncut disc = πr^2
$$= 3.142 × 65 × 65$$
$$= 13\,275 \text{ mm}^2$$
Volume of uncut disc = 13 275 × 20
$$= 265\,500 \text{ mm}^3$$
Volume of finished flange = 265 500 − 72 648
$$= 192\,852 \text{ mm}^3$$
$$= 192\,900 \text{ mm}^3 \quad (Ans.)$$

6.7 Surface Area

1 Sometimes it is required to find the total **surface area** of a solid, i.e. the sum of the areas of all its surfaces. This information is necessary for such operations as electro-plating or other surface-coating processes.

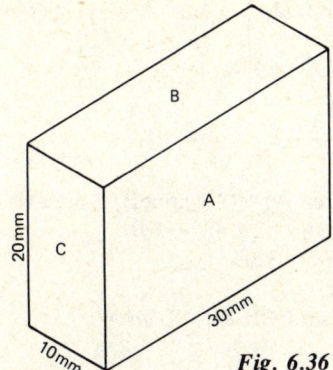

Fig. 6.36

The block of metal shown in fig. 6.36 is a rectangular prism. The block has 6 surfaces consisting of

2 side faces (marked A) each 30 mm × 20 mm
1 top and 1 bottom face (marked B) each 30 mm × 10 mm
2 end faces (marked C) each 10 mm × 20 mm

∴ Total surface area of the block is given by

$$
\begin{array}{r}
2 \times 30 \times 20 = 1\,200 \\
+ 2 \times 30 \times 10 = 600 \\
+ 2 \times 10 \times 20 = \underline{400} \\
2\,200
\end{array}
$$

Surface area of the block = 2 200 mm²

Example 6.34 Calculate the total surface area of a rectangular block
16 cm × 8 cm × 4 cm
The block is shown in fig. 6.37.

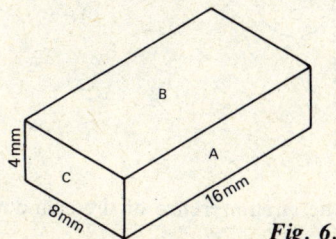

Fig. 6.37

$$
\begin{array}{rl}
\text{Total surface area} = & \text{(A) } 2 \times 16 \times 4 = 128 \\
& \text{(B) } 2 \times 16 \times 8 = 256 \\
& \text{(C) } 2 \times 8 \times 4 = \underline{64} \\
& \phantom{\text{(C) } 2 \times 8 \times 4 = 1}448
\end{array}
$$

Total surface area = 448 cm²   (*Ans.*)

Example 6.35 Calculate the total surface area of a cube 20 mm side.
The block is shown in fig. 6.38.

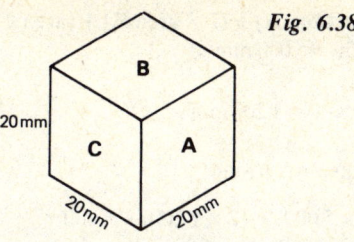

Fig. 6.38

The cube has 6 equal sides (2A + 2B + 2C)
Each side has an area = 20 × 20 = 400 mm²
∴   Total surface area = 6 × 400
$$= 2\,400 \text{ mm}^2 \quad (\textit{Ans.})$$

2 To find the total surface area of some solids, it is useful to imagine that the solid is made from cardboard which can be opened up and laid out flat. Then the area of the flat cardboard is equal to the total surface area of the solid.

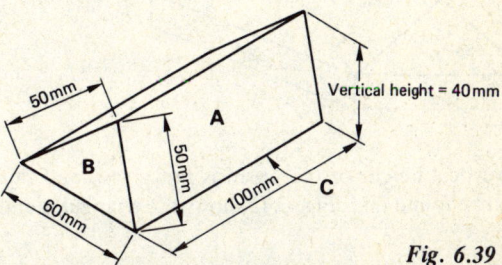

Fig. 6.39

Fig. 6.39 shows a triangular prism. The prism has 2 equal rectangular sides (marked A), 2 equal triangular ends (marked B), and a rectangular base (marked C). Imagine the prism has been split open along its top edge and laid out flat (fig. 6.40).

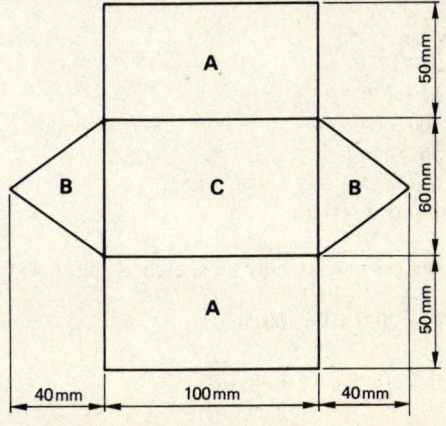

Fig. 6.40

This flat shape is called the **net** of the triangular prism.

Area of the net = total surface area of the prism

From fig. 6.40:
Area of net = (2 × area A) + (2 × area B) + (area C)
Area A = 100 × 50 = 5 000 mm²

Area B = $\frac{1}{2}$ × 60 × 40 = 1 200 mm²

Area C = 100 × 60 = 6 000 mm²

Area of net = (2 × 5 000) + (2 × 1 200) + 6 000
= 10 000 + 2 400 + 6 000
= 18 400 mm²

Surface area of prism = 18 400 mm²   (*Ans.*)

Example 6.36   Calculate the total surface area of the triangular prism shown in fig. 6.41.

Fig. 6.41

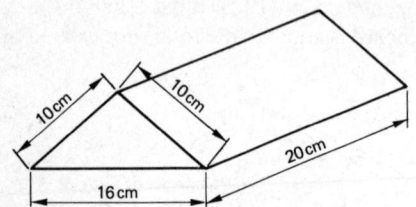

The vertical height of the prism is not given, and this should be found first. Fig. 6.42 shows the triangular end face.

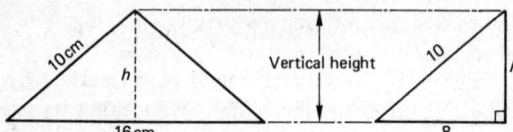

Fig. 6.42

By Pythagoras:
$10^2 = 8^2 + h^2$
$100 = 64 + h^2$
$h^2 = 100 - 64$
$h^2 = 36$
$h = 6$
∴   Vertical height = 6 cm

The net of the prism may now be sketched (fig. 6.43).

Area of A = 20 × 10 = 200 cm²

Area of B = $\frac{1}{2}$ × 16 × 6 = 48 cm²

Area of C = 20 × 16 = 320 cm²

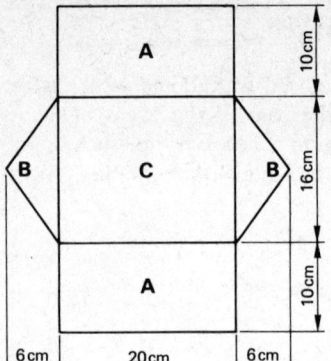

Fig. 6.43

Area of net = (2 × area A) + (2 × area B) + area C
= (2 × 200) + (2 × 48) + 320
= 400 + 96 + 320
= 816 cm²

Surface area of the prism = 816 cm²   (*Ans.*)

3 Fig. 6.44 shows a **solid cylinder** of radius 7 cm and length 20 cm. An **imaginary net** can be drawn for the cylinder by cutting along the length and circumference, leaving the end faces as attached circles.

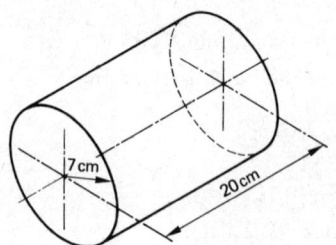

Fig. 6.44

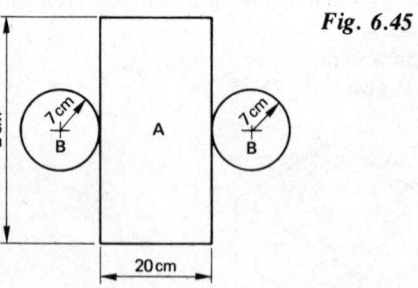

Fig. 6.45

The dimension L is the circumference of the cylinder and is given by

$L = 2\pi r$

$= 2 \times \frac{22}{7} \times 7$ when $r = 7$ cm

$= 44$ cm

Area of net
= area of rectangle A + 2 × area of circle B
Area of rectangle A = 20 × 44 = 880 cm²
Area of circle B = πr^2

$$= \frac{22}{7} \times 7 \times 7 = 154\,cm^2$$

Area of net = 880 + (2 × 154)
= 880 + 308 = 1 188 cm²
Total surface area of cylinder = 1 188 cm²

Example 6.37 Find the curved surface area of the open-ended tube shown in fig. 6.46.

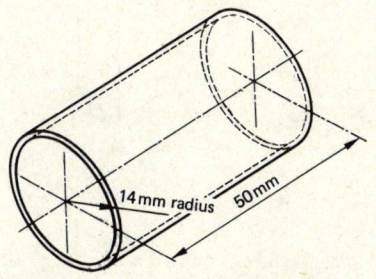

Fig. 6.46

The tube has no end faces so that its total surface area equals the area of the curved surface. Drawing the net (fig. 6.47),

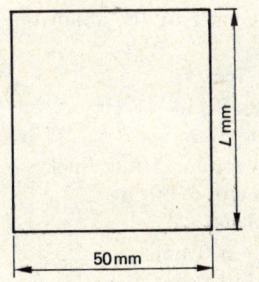

Fig. 6.47

$L = 2\pi r$ where $r = 14$ mm

$= 2 \times \dfrac{22}{7} \times 14 = 88$ mm

Area of net = area of rectangle
= 50 × 88 = 4 400 mm²
Curved surface area = 4 400 mm²   (*Ans.*)

4 The total surface area of a **pyramid** may be found in the same way. A net for a square pyramid is shown in fig. 6.48.

Total surface area of pyramid
= area of rectangle A + (4 × area of triangle B)

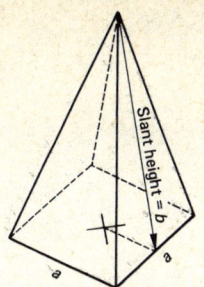

Fig. 6.48

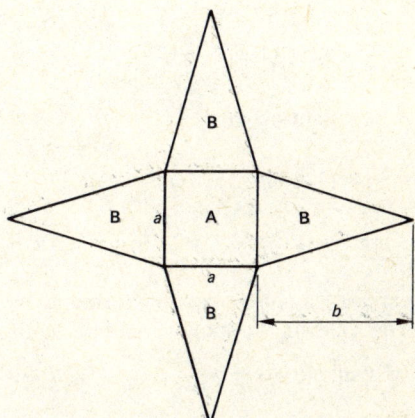

5 The use of a net to find the total surface area of a **cone** is too difficult at this stage. The following formula may be used, however:

Total surface area of cone
= area of circular base + curved area

Area of circular base = πr^2
Curved area = $\pi r \times$ slant height

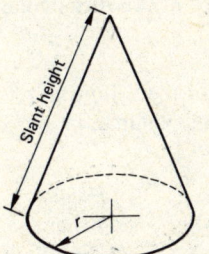

Fig. 6.49

Example 6.38 Calculate the total surface area of the cone shown in fig. 6.50.

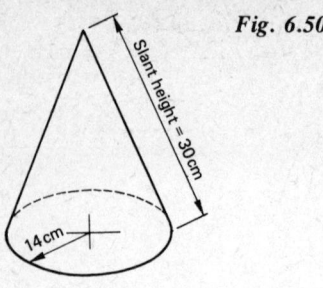

Fig. 6.50

Slant height = 30 cm

14 cm

Area of base $= \pi r^2$

$$= \frac{22}{7} \times 14 \times 14 = 616\, \text{cm}^2$$

Curved area $= \pi r \times$ slant height

$$= \frac{22}{7} \times 14 \times 30 = 1\,320\, \text{cm}^2$$

Total surface area $= 616 + 1\,320 = 1\,936\, \text{cm}^2$ (*Ans.*)

6 The surface area of a **sphere** cannot be found by drawing a net. The following formula should be used:

Surface area of a sphere $= 4\pi r^2$

Example 6.39 Calculate the surface area of a sphere 36 mm in diameter. Take $\pi = 3.142$.

Radius of sphere $= \dfrac{36}{2} = 18\, \text{mm}$

Surface area $= 4\pi r^2$
$$= 4 \times 3.142 \times 18 \times 18$$
$$= 4\,072\, \text{mm}^2 \quad (\textit{Ans.})$$

Exercises 6

6.1 Determine the volume of each of the rectangular prisms shown in fig. 6.51.

6.2 Sketch and dimension each of the following rectangular steel blocks and find its volume in cubic millimetres:

	Length	Width	Height
a)	200 mm	30 mm	40 mm
b)	85 mm	17 mm	6 mm
c)	64 mm	20.5 mm	18 mm
d)	28 mm	12.5 mm	9.8 mm
e)	30 cm	15 mm	40 mm
f)	0.5 m	25 mm	25 mm
g)	1.2 m	50 mm	37 mm
h)	19 mm	4.5 mm	2.6 mm

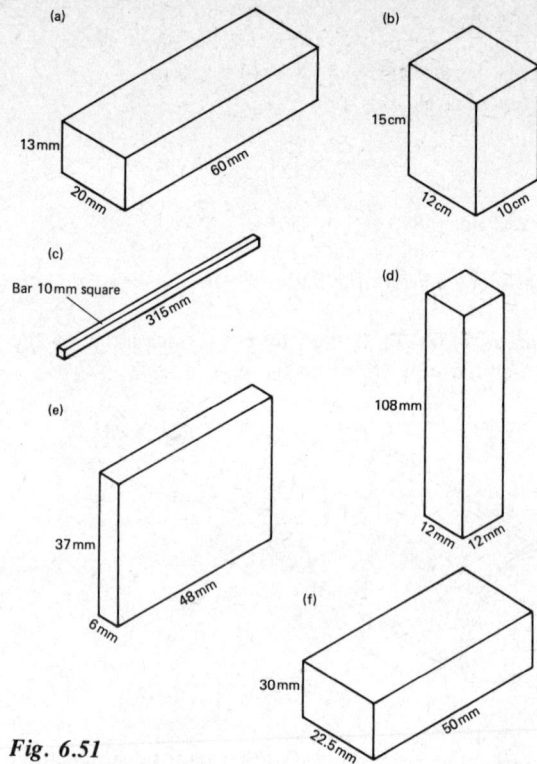

Fig. 6.51

(a) 13 mm, 20 mm, 60 mm

(b) 15 cm, 12 cm, 10 cm

(c) Bar 10 mm square, 315 mm

(d) 108 mm, 12 mm, 12 mm

(e) 37 mm, 6 mm, 48 mm

(f) 30 mm, 22.5 mm, 50 mm

6.3 Calculate the volume of each of the following rectangular solids:
a) 2 m length of 20 mm square bar
b) concrete machine foundation 2.1 m × 1.4 m × 0.5 m
c) $1\frac{1}{2}$ m length of timber 4 cm by 6 cm
d) steel levelling plate 250 mm square 20 mm thick
e) cast iron block 40 mm × 80 mm × 90 mm
f) sheet metal panel 1 m × 350 mm × 4 mm
g) steel billet 200 mm square × 600 mm
h) paving stone 45 cm square and 8 cm thick.

6.4 The sizes of a set of precision steel blocks are given below. Calculate the volume of each block to the nearest cubic millimetre.

	Length	Width	Height
a)	35 mm	25 mm	5.50 mm
b)	35 mm	25 mm	5.85 mm
c)	35 mm	25 mm	7.35 mm
d)	35 mm	25 mm	10.15 mm

6.5 Calculate the volume of each of the prisms shown in fig. 6.52.

6.6 A firm produces a range of aluminium extrusions of the cross-sections shown in fig. 6.53. Calculate the volume of a length of 800 mm of each cross-section.

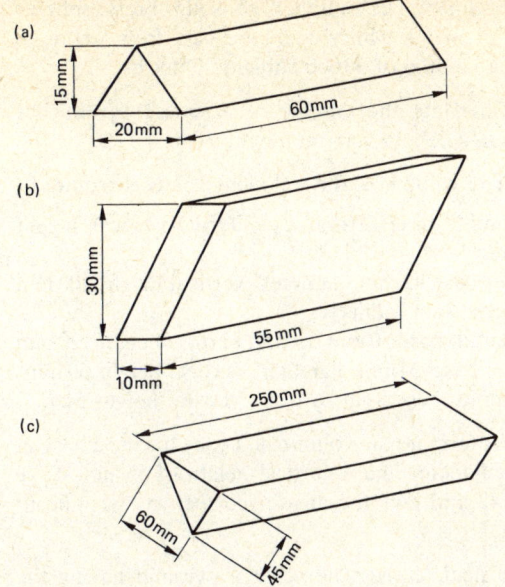

Fig. 6.52

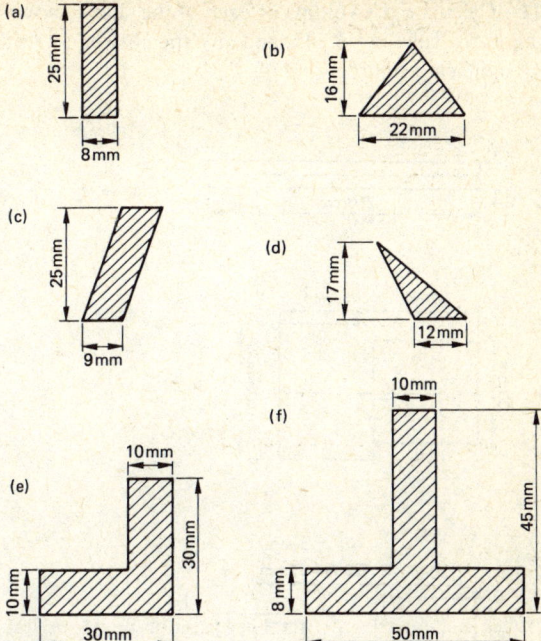

Fig. 6.53

6.7 Calculate the volume of a 2.2 m length of a beam having the cross-section shown in fig. 6.54.

6.8 Calculate the volume of the shear blade shown in fig. 6.55.

6.9 Calculate the volume of the following cylinders; take $\pi = \frac{22}{7}$.
a) 14 mm diameter by 60 mm long
b) 7 cm diameter by 10 cm long
c) 21 mm radius by 100 mm long
d) 1.4 m diameter by 2 m long
e) 70 mm diameter by 250 mm long
f) 42 mm diameter by 140 mm long.

6.10 The volume of the following cylinders is required. Take $\pi = 3.142$ and give the answer correct to 3 significant figures.
a) 12.5 mm diameter by 44 mm long
b) 8.3 cm radius by 20 cm long
c) 0.75 m radius by 1.2 m long
d) 45 mm diameter by 85 mm long
e) 7.5 mm diameter by 35 mm long
f) 0.5 mm radius by 6.4 mm long.

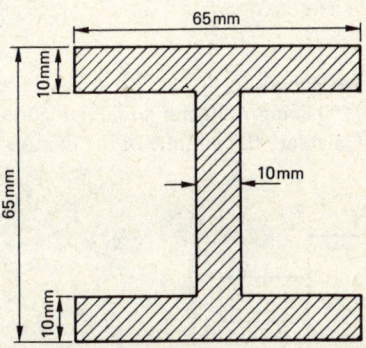

Fig. 6.54

Fig. 6.55

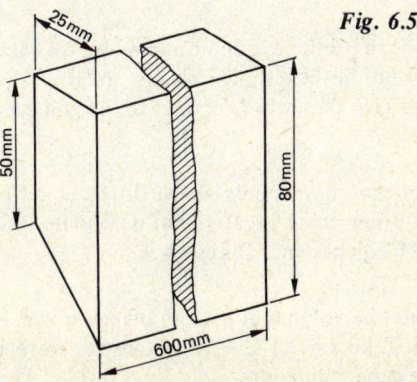

Volume and Surface Area 69

6.11 Calculate the volume of each of the shafts shown in fig. 6.56. Take $\pi = 3.142$, and give the answer correct to 3 significant figures.

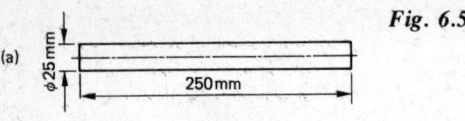

(a)

Fig. 6.56

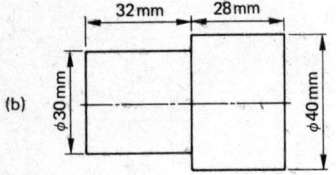

(b)

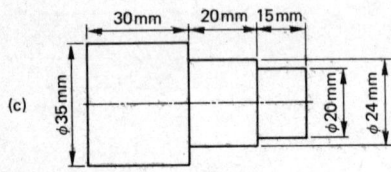

(c)

6.12 The component shown in fig. 6.57 consists of a 28 mm diameter shaft having a 10 mm square portion milled on one end. Calculate the volume of the component. Take $\pi = \frac{22}{7}$.

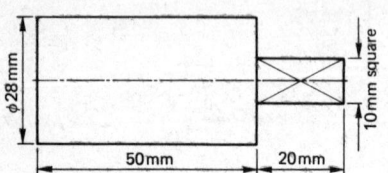

Fig. 6.57

6.13 Calculate the volume in cubic millimetres of each of the cast iron bushes shown in fig. 6.58. Take $\pi = 3.142$, and give the answer correct to 3 significant figures.

6.14 Each of the components shown in fig. 6.59 is made from 10 mm thick steel plate. Calculate the volume of each component. Take $\pi = \frac{22}{7}$.

6.15 Calculate the volume of a 6 mm diameter precision steel ball. Take $\pi = 3.142$; give the answer correct to the nearest cubic millimetre.

6.16 Estimate the number of solid brass spheres 20 mm diameter which could be cast from an ingot having a volume of 30 000 cubic millimetres.

6.17 Calculate the volume of a square pyramid of base 5 cm side and vertical height 9 cm.

6.18 The volume of the following solids is required:

a) pyramid: base 20 mm by 30 mm, vertical height 40 mm
b) cone: base 42 mm diameter, vertical height 100 mm
c) sphere: 7 cm radius
d) pyramid: base 50 mm square, vertical height 120 mm
e) cone: base 20 mm diameter, vertical height 60 mm
f) pyramid: base 8 cm by 6 cm, vertical height 5 cm.

6.19 Determine the volume of a cone having a base of 30 mm diameter and a vertical height of 55 mm. Take $\pi = 3.142$ and give the answer correct to 3 significant figures.

6.20 Calculate the volume of a pyramid having the base shown in fig. 6.60 and a vertical height of 75 mm.

6.21 Determine the volume in cubic metres that could be contained in the following rectangular packing crates. The internal dimensions of the crates are given in metres.

	Length	*Width*	*Height*
a)	2	1	1.5
b)	1.8	1.5	1
c)	2.2	1.4	0.5
d)	1.7	0.8	0.6
e)	2	1.4	0.75
f)	2.5	1.25	1.65
g)	0.7	0.45	0.3
h)	3	0.95	0.4

6.22 Calculate the volume to the nearest cubic centimetre that could be contained in the sheet metal box shown in fig. 6.61. (Neglect the thickness of the sheet metal.)

6.23 Calculate the capacity of the following containers. Give the answer to the nearest litre.

a) oil drum: 60 cm diameter by 75 cm height
b) rectangular coolant tank: 600 mm × 300 mm × 150 mm
c) hydraulic cylinder: 100 mm diameter × 130 mm long
d) rectangular oil cooler: 250 mm × 150 mm × 80 mm
e) emergency water tank: 3.2 m × 2.1 m × 1.8 m
f) fire extinguisher: 21 cm diameter × 60 cm.

6.24 A workpiece 30 mm diameter and 200 mm long is reduced to 20 mm diameter throughout its length in a turning operation. Determine the volume of metal removed in machining.

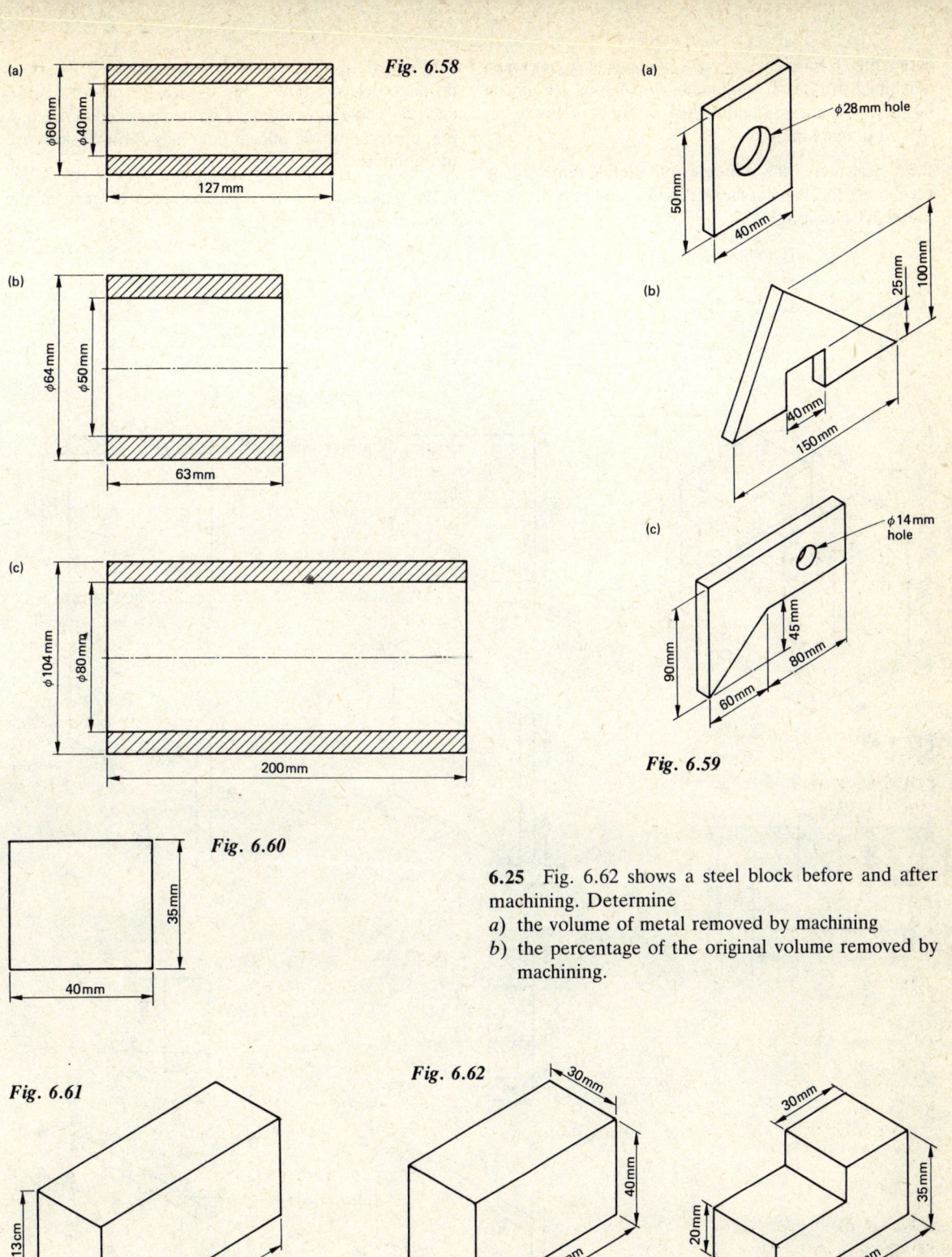

Fig. 6.58

Fig. 6.59

Fig. 6.60

Fig. 6.61

6.25 Fig. 6.62 shows a steel block before and after machining. Determine

a) the volume of metal removed by machining

b) the percentage of the original volume removed by machining.

Fig. 6.62

BEFORE MACHINING

AFTER MACHINING

6.26 Fig. 6.63 shows a component which has been machined from a rectangular block of steel 100 mm × 60 mm × 80 mm. Calculate the volume of metal removed in machining.

6.27 Calculate the volume of metal removed in machining the tee-slot shown in fig. 6.64 in a block of metal 400 mm long.

6.28 In fig. 6.65 the component marked B has been produced by machining the workpiece marked A. In each case give the volume of metal removed in machining, correct to 3 significant figures. All dimensions are in millimetres.

6.29 Calculate the total surface area of each of the solids shown in fig. 6.66.

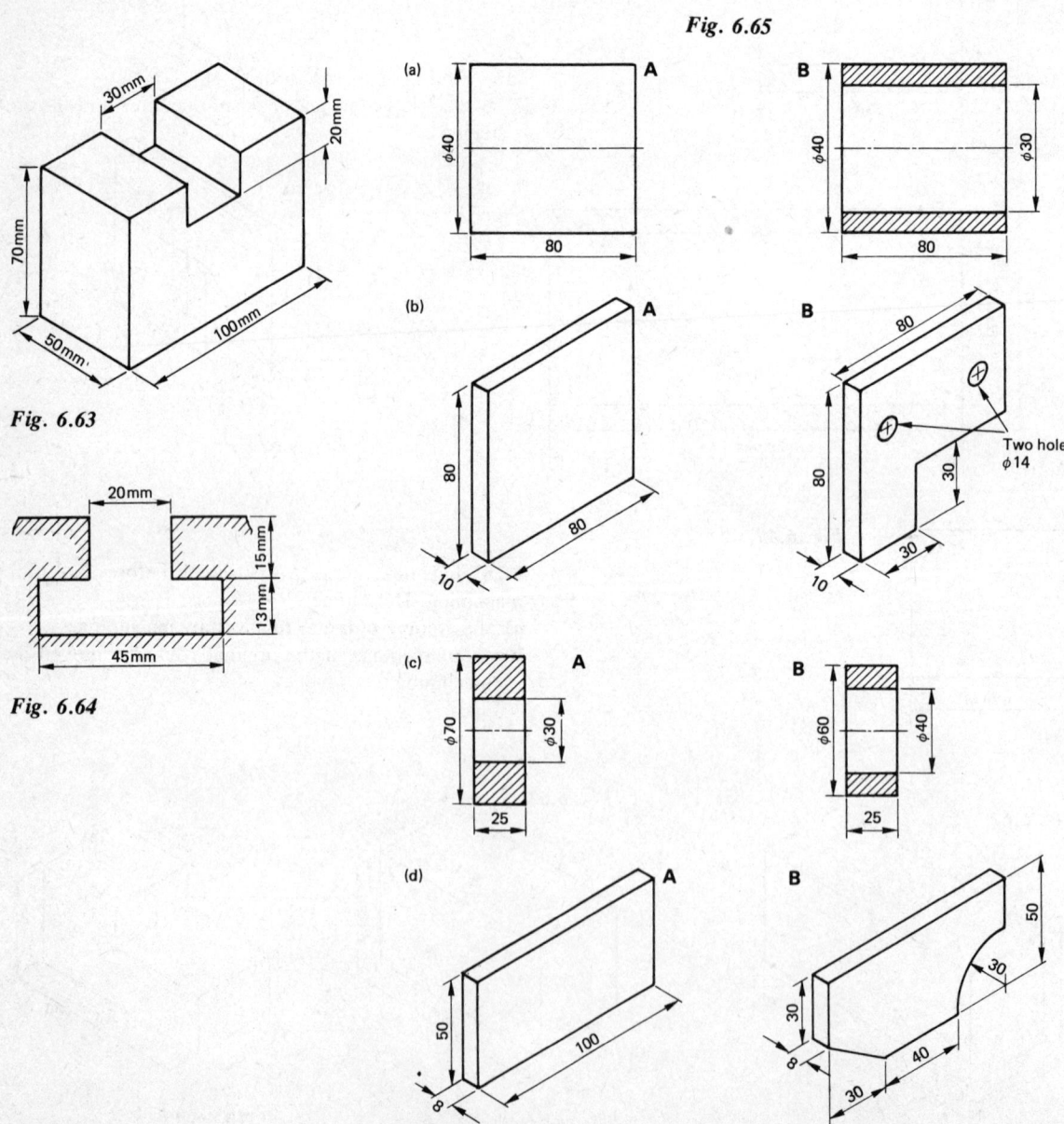

Fig. 6.65

Fig. 6.63

Fig. 6.64

6.30 Calculate
a) the curved surface area of an open-ended tube of 42 mm diameter and 80 mm length
b) the surface area of a sphere 20 mm diameter
c) the total surface area of a cone having a base of radius 10 cm and slant height of 25 cm.

6.31 A rectangular bus-bar has the dimensions 20 mm × 15 mm × 3 m long. Find a) its volume in cubic millimetres, b) its surface area in square millimetres.

6.32 A packing case for a transformer has the following dimensions:
External, m   1 × 1.5 × 2
Internal, m   0.95 × 1.45 × 1.95
Calculate
a) the volume occupied by the packing case for shipping purposes
b) the volume contained by the packing case.

Fig. 6.66

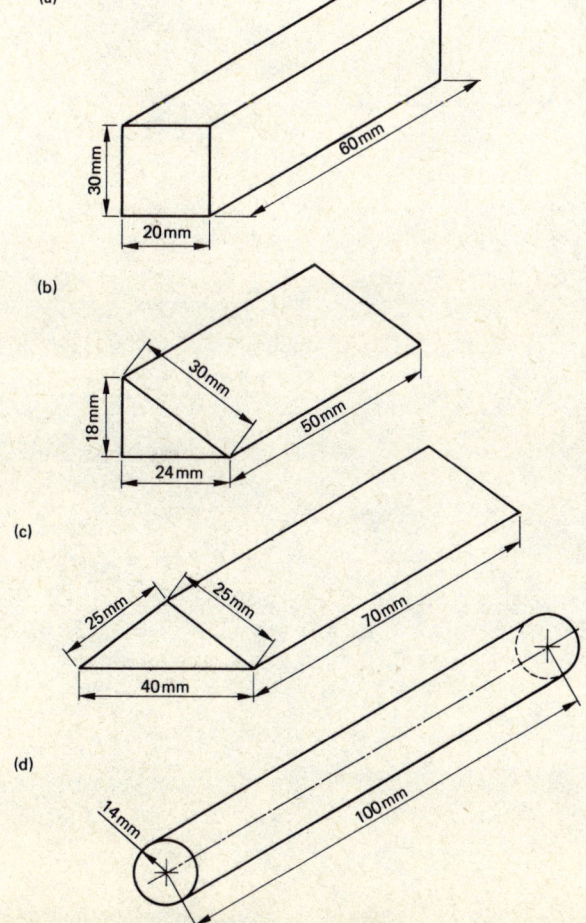

6.33 How many litres of transformer oil could be contained in a drum of internal dimensions 0.7 m diameter and 0.9 m height?

6.34 The table shows the external dimensions of three types of 1.5 V dry battery. Complete the table by calculating the volume of each battery.

TYPE	DIAMETER (mm)	LENGTH (mm)	VOLUME (mm³)
HP 2	32	58	
HP 7	13	48	
HP 11	25	46	

6.35 Find the capacity in litres of
a) an electric water heater 0.4 m diameter and 0.6 m height
b) an oil dashpot 50 mm diameter by 75 mm
c) an electric kettle 18 cm diameter by 14 cm.
(Give the answers to 3 s.f.)

6.36 Find the volume in cubic millimetres of
a) an earthing rod 15 mm diameter by 420 mm long
b) a length of aluminium cable 2 mm diameter by 2 m long
c) an armature 80 mm diameter by 250 mm long.
(Give the answers to 4 s.f.)

6.37 Find the volume contained in cubic millimetres by
a) a 1.6 m length of metal trunking of internal dimensions 130 mm by 65 mm
b) an electrician's toolbox of internal dimensions 200 mm × 320 mm × 450 mm.

6.38 Calculate the volume of the permanent magnet shown in fig. 6.67.

Fig. 6.67

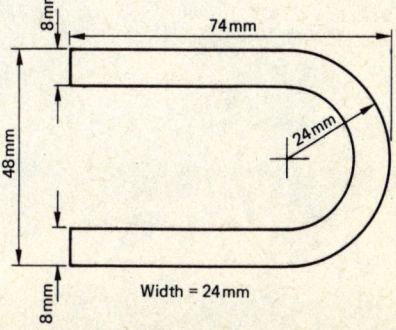

6.39 Calculate the capacity in litres of the immersion heater storage tank shown in fig. 6.68.

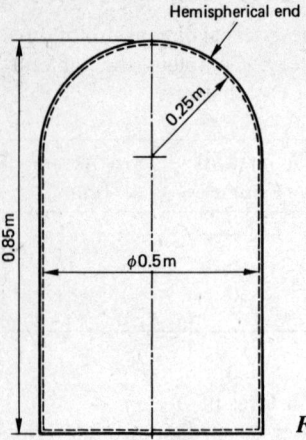

Hemispherical end

0.25 m

0.85 m

φ0.5 m

Fig. 6.68

6.40 Find the volume of a bar magnet 8 mm × 4.5 mm × 135 mm.

7 Angles

Frequently, workshop calculations are concerned with angular dimensions as well as length dimensions; e.g. machining settings for taper shafts, grinding tool angles, checking engine timing. Various types of measuring instrument are used by craftsmen for the precise measurement of angle. These include: the bevel protractor, vernier protractor, clinometer, optical dividing head, angle dekkor, and auto-collimator.

It is important to be able to convert a required angle into units corresponding to the scales of the measuring instruments.

7.1 Units of Angle

Angle is measured as the rotation in a circle. The angle x in fig. 7.1 is the amount of rotation of the line OA to reach the position OB.

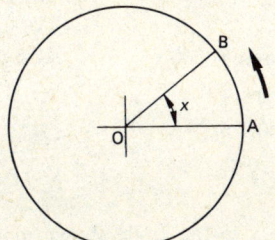

Fig. 7.1

The basic **unit of angle** is the **degree** and is defined as $\frac{1}{360}$th of one complete rotation of a circle.

Hence 360 degrees = 1 revolution
usually written as 360° = 1 rev

The degree is divided into 60 parts to give **minutes** of angle.

$$1 \text{ degree} = 60 \text{ minutes}$$
usually written as $1° = 60'$

The minute is divided into 60 parts to give **seconds** of angle.

$$1 \text{ minute} = 60 \text{ seconds}$$
usually written as $1' = 60''$

Thus an angle of 16 degrees 42 minutes 28 seconds would be written as

$$16° \ 42' \ 28''$$

To convert degrees to minutes we multiply by 60.
degrees × 60 = minutes

$$\therefore \quad \frac{\text{minutes}}{60} = \text{degrees}$$

To convert minutes to seconds we multiply by 60.
minutes × 60 = seconds

$$\therefore \quad \frac{\text{seconds}}{60} = \text{minutes}$$

Example 7.1 Convert the angle 4° 35′ to minutes.

$$\begin{aligned} \text{Angle in minutes} &= (4 \times 60) + 35 \\ &= 240 + 35 = 275' \quad (Ans.) \end{aligned}$$

Example 7.2 Convert 5° 24′ to seconds.

$$\begin{aligned} \text{Angle in minutes} &= (5 \times 60) + 24 \\ &= 300 + 24 = 324' \\ \text{Angle in seconds} &= 324 \times 60 = 19\,440'' \quad (Ans.) \end{aligned}$$

Example 7.3 Convert 7 000 seconds to an angle in degrees, minutes and seconds.

Converting seconds to minutes:

$$\begin{aligned} \frac{7\,000}{60} &= 116 \text{ remainder 40 seconds} \\ &= 116' \ 40'' \end{aligned}$$

Converting minutes to degrees:

$$\begin{aligned} \frac{116}{60} &= 1 \text{ remainder 56 minutes} \\ &= 1° \ 56' \end{aligned}$$

$$\therefore \quad \text{Angle} = 1° \ 56' \ 40'' \quad (Ans.)$$

Sometimes an angle may be given or required as a **decimal of a degree**.

Example 7.4 Convert 17.34 degrees to an angle in degrees, minutes and seconds.

The 17 degrees remains unchanged.
To convert the decimal portion to minutes:
   $0.34 \times 60 = 20.4'$

To convert the decimal minutes to seconds:
   $0.4 \times 60 = 24''$

∴   Angle = 17° 20′ 24″   *(Ans.)*

Example 7.5   Convert the angle 18° 24′ 36″ to decimal.

The 18 degrees remains unchanged.
First convert the seconds to minutes:

$$\frac{36}{60} = 0.6'$$

Thus   24′ 36″ = 24.6′

Now convert the minutes to degrees:

$$\frac{24.6}{60} = 0.41°$$

∴   Angle = 18.41°   *(Ans.)*

Example 7.6   The following angle measurements were taken in a test using a precision clinometer:
   22° 41′   23° 04′   22° 36′   21° 59′
Find the average of the four measurements.

   Average measurement

$$= \frac{\text{sum of the measurements}}{\text{number of measurements}}$$

$$= \frac{22° \ 41' + 23° \ 04' + 22° \ 36' + 21° \ 59'}{4}$$

$$= \frac{90° \ 20'}{4}$$

$$= 22° \ 35' \quad (Ans.)$$

7.2 Angles in Circle and Semi-circle

If the line OA in fig. 7.2 makes one complete revolution, it will rotate through an angle of 360°.

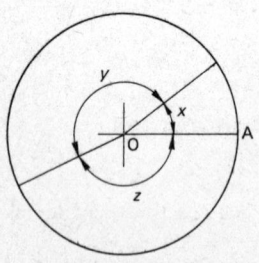

Fig. 7.2

Hence $x + y + z = 360°$

This gives the general expression:

   The sum of the angles in a **circle** = 360°

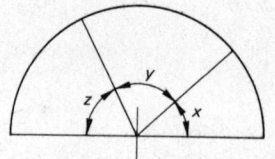

Fig. 7.3

The semi-circle shown in fig. 7.3 is half a circle:

   $x + y + z = 180°$

   The sum of the angles in a **semi-circle** = 180°

Example 7.7   Find the angle x in the circle shown in fig. 7.4.

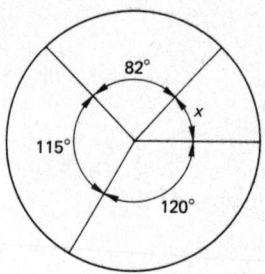

Fig. 7.4

The sum of the angles in a circle = 360°

∴ $x + 82° + 115° + 120° = 360°$
   $x + 317° = 360°$

$x = 360° - 317° = 43°$ *(Ans.)*

Example 7.8   Find the angle x in the circle shown in fig. 7.5.

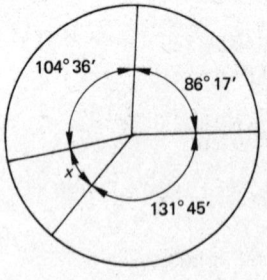

Fig. 7.5

The sum of the angles in a circle = 360°

∴ $x + 131° \ 45' + 86° \ 17' + 104° \ 36' = 360°$
   $x + 322° \ 38' = 360°$

$x = 360° - 322° \ 38' = 37° \ 22'$ *(Ans.)*

Example 7.9 Find the angle *x* in the semi-circle shown in fig. 7.6.

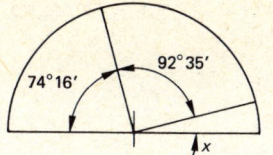

Fig. 7.6

Sum of the angles in a semi-circle = 180°

$$\therefore \quad x + 92° \ 35' + 74° \ 16' = 180°$$
$$x + 166° \ 51' = 180°$$

$$x = 180° - 166° \ 51' = 13° \ 09' \quad (Ans.)$$

7.3 The Right-angle

As line OA in fig. 7.7 rotates to position OB, it moves through one quarter of a revolution.

One revolution = 360°
$\frac{1}{4}$ revolution = 90°

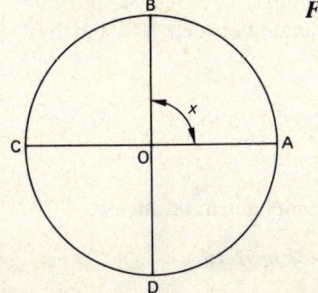

Fig. 7.7

It can be seen from the diagram that OB is *perpendicular* or "square" to OA. Thus, the angle between a horizontal line and a perpendicular is 90° and is given the name **right-angle**. In an engineer's square, for instance, the perpendicular blade is said to be at right-angles to the base.

From fig. 7.7 it can be seen that there are four right-angles in a full circle.

$$\therefore \quad \text{Circle} = 360° = 4 \text{ right-angles}$$

A right-angle may be indicated on a diagram in either of the ways shown on the right-angled triangles in fig. 7.8.

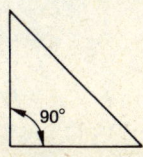 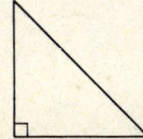

Fig. 7.8

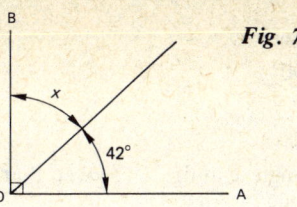

Fig. 7.9

In fig. 7.9 the line OB is at right-angles to the line OA. The sum of the angles in a right-angle is 90°.

$$\therefore \quad x + 42° = 90°$$
$$x = 90° - 42°$$
$$= 48°$$

Example 7.10 Find the angle *x* in fig. 7.10.

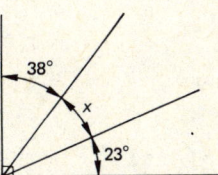

Fig. 7.10

Sum of the angles in a right-angle = 90°

$$\therefore \quad x + 38° + 23° = 90°$$
$$x + 61° = 90°$$

$$x = 90° - 61° = 29° \quad (Ans.)$$

Example 7.11 Determine the rake angle of the cutting tool shown in fig. 7.11.

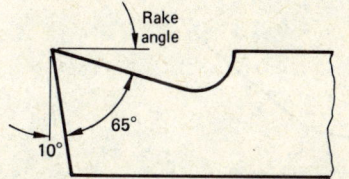

Fig. 7.11

Sum of the angles in a right-angle = 90°

$$\therefore \quad \text{Rake angle} + 65° + 10° = 90°$$
$$\text{Rake angle} + 75° = 90°$$

$$\text{Rake angle} = 90° - 75° = 15° \quad (Ans.)$$

Example 7.12 Find the angle *x* in fig. 7.12.

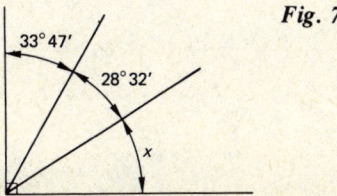

Fig. 7.12

$$x + 28° 32' + 33° 47' = 90°$$
$$x + 62° 19' = 90°$$
$$x = 90° - 62° 19' = 27° 41' \quad (Ans.)$$

Example 7.13 Find the angle x in fig. 7.13 given that the full angle is a right-angle.

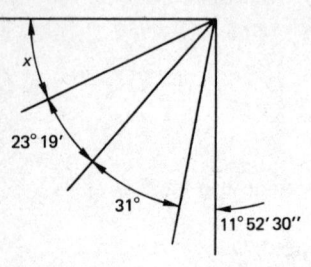

Fig. 7.13

$$x + 23° 19' + 31° + 11° 52' 30'' = 90°$$
$$x + 66° 11' 30'' = 90°$$
$$x = 90° - 66° 11' 30'' = 23° 48' 30'' \quad (Ans.)$$

Example 7.14 Fig. 7.14 shows the angles at the cutting point of a chisel. Calculate the value of the rake angle when the point angle is 60° and the clearance angle is 10°.

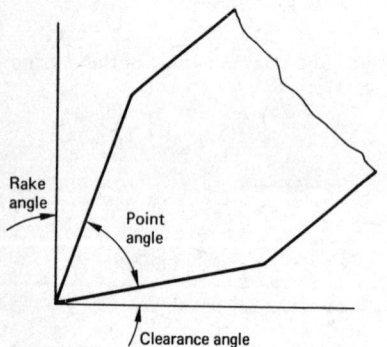

Fig. 7.14

Rake angle + point angle + clearance angle = 90°

$$\text{Rake angle} + 60° + 10° = 90°$$
$$\text{Rake angle} + 70° = 90°$$

$$\text{Rake angle} = 90° - 70° = 20° \quad (Ans.)$$

7.4 Types of Angle

Fig. 7.15 shows the names which are given to angles depending on their magnitude.

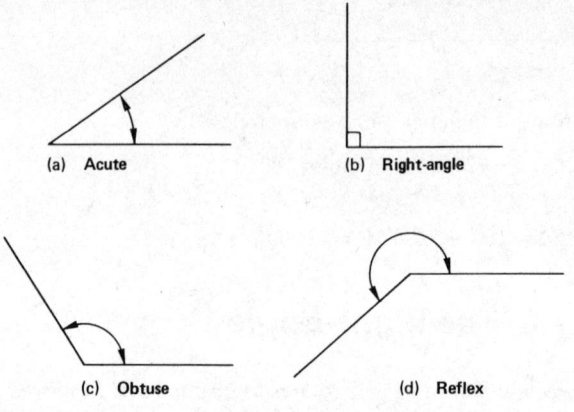

Fig. 7.15

a) An **acute angle** is any angle less than 90°.
b) A **right-angle** is an angle of 90°.
c) An **obtuse angle** is any angle between 90° and 180°.
d) A **reflex angle** is any angle between 180° and 360°.

1 The angles in fig. 7.16 add up to 90°

$$x + y = 90°$$

These angles are called **complementary angles**.

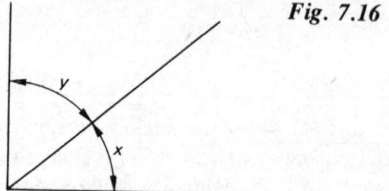

Fig. 7.16

2 The angles in fig. 7.17 add up to 180°

$$x + y = 180°$$

These angles are called **supplementary angles**.

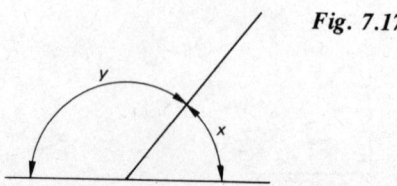

Fig. 7.17

Example 7.15   Find the angle complementary to 63° 14'.

Let x = angle to be found.
Complementary angles add up to 90°.

$\therefore$ $x + 63°\ 14' = 90°$

$x = 90° - 63°\ 14' = 26°\ 46'$ (*Ans.*)

Example 7.16   Find the angle supplementary to 58° 35'.

Let x = angle to be found.
Supplementary angles add up to 180°.

$\therefore$ $x + 58°\ 35' = 180°$

$x = 180° - 58°\ 35' = 121°\ 25'$ (*Ans.*)

7.5 Angles and Straight Lines

1 Fig. 7.18 shows the intersection of two straight lines. The angles on opposite sides of the point of inter-

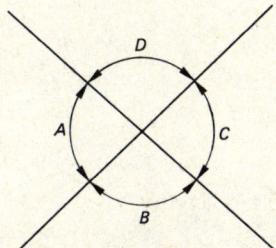

Fig. 7.18

section are equal. These angles are called **vertically opposite angles**.
   Thus, vertically opposite angles are equal.

   Angle A = angle C
   Angle B = angle D

Angle A and angle B are supplementary angles.
$\therefore$ angle A + angle B = 180°
Angle C and angle D are supplementary angles.
$\therefore$ angle C + angle D = 180°

In fig. 7.19, the angle x is opposite to the known angle of 85°. Therefore x and 85° are vertically opposite angles and are equal.
$\therefore$ $x = 85°$

Angle z and 85° are supplementary angles.
$\therefore$ $z + 85° = 180°$
   $z = 180° - 85° = 95°$

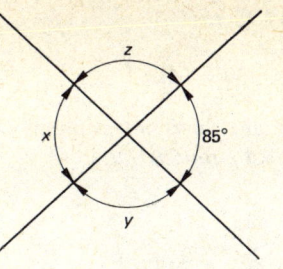

Fig. 7.19

Angle y and angle z are vertically opposite angles.
$\therefore$ $y = z = 95°$

To check these results, the sum of the four angles should give a full circle about the point of intersection, i.e. 360°.
$\therefore$ $x + y + z + 85° = 360°$
   $85° + 95° + 95° + 85° = 360°$

Example 7.17   Determine the angles A, B and C in fig. 7.20.

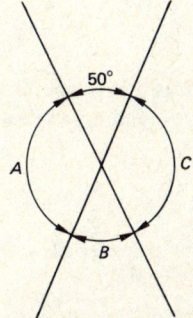

Fig. 7.20

Angles B and 50° are vertically opposite angles.
$\therefore$ $B = 50°$
Angles C and 50° are supplementary angles.
$\therefore$ $C + 50° = 180°$
   $C = 180° - 50° = 130°$
Angle A and angle C are vertically opposite angles.
$\therefore$ $A = C = 130°$
Insert the angles (fig. 7.21) and check:

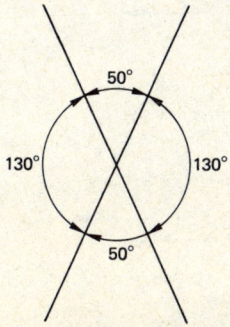

Fig. 7.21

$$A + B + C + 50° = 360°$$
$$130° + 50° + 130° + 50° = 360°$$
$$\therefore \quad A = 130° \quad B = 50° \quad C = 130° \quad (Ans.)$$

2 Fig. 7.22 shows two parallel lines cut at points O and P by a straight line called a *transversal*.

Fig. 7.22

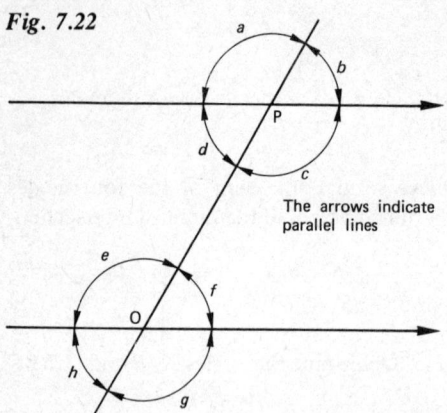

The arrows indicate parallel lines

For each angle about point P, there is a corresponding angle about point O. These corresponding angles are equal.

Thus $a = e$ $b = f$
       $c = g$ $d = h$

The angles on alternate sides of the transversal are called **alternate angles** and are equal. That is
$$d = f \quad c = e$$

There is also a number of pairs of supplementary angles:
$$a + b = 180° \qquad f + g = 180°$$

Example 7.18   Find all the missing angles in fig. 7.23.

Fig. 7.23

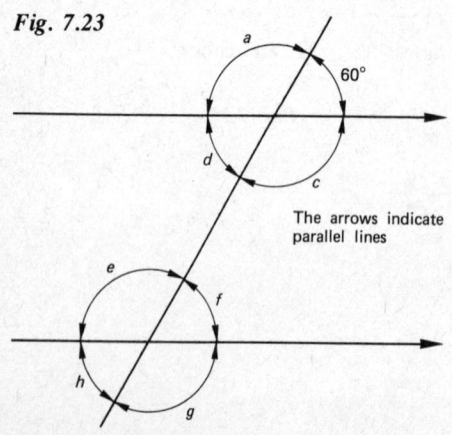

The arrows indicate parallel lines

Angle a and 60° are supplementary angles.
$$\therefore \quad a + 60° = 180°$$
$$a = 180° - 60° = 120°$$
Angle a and angle c are vertically opposite angles.
$$\therefore \quad c = a = 120°$$
Angle d and 60° are vertically opposite angles.
$$\therefore \quad d = 60°$$
The following are corresponding angles:
$$\begin{aligned} a &= e & \therefore \quad e &= 120° \\ 60° &= f & \therefore \quad f &= 60° \\ c &= g & \therefore \quad g &= 120° \\ d &= h & \therefore \quad h &= 60° \end{aligned}$$
Therefore, the missing angles are
$$a = 120° \quad c = 120° \quad d = 60° \quad e = 120°$$
$$f = 60° \quad g = 120° \quad h = 60° \quad (Ans.)$$

Example 7.19   Determine the angles x and y in fig. 7.24

Fig. 7.24

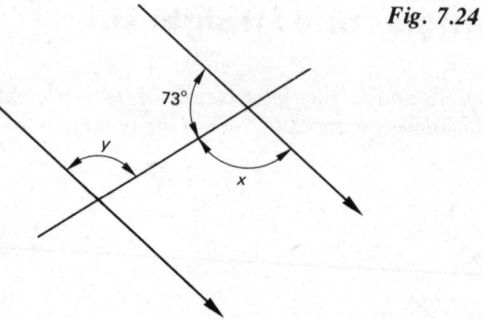

$$x + 73° = 180° \quad \text{(supplementary angles)}$$
$$x = 180° - 73° = 107°$$

$$y = x \quad \text{(alternate angles)}$$
$$= 107°$$

Therefore $x = 107°$ and $y = 107°$ (*Ans.*)

7.6 Angles in a Triangle

1   The angles A, B and C in the triangle shown in fig. 7.25 are called **interior angles**.

Fig. 7.25

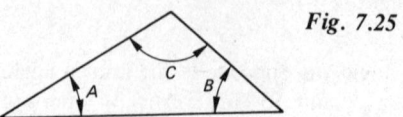

The sum of the interior angles of any triangle is 180°.
$$\therefore \quad A + B + C = 180°$$
This relationship can be used to find the missing angle in the triangle shown in fig. 7.26.

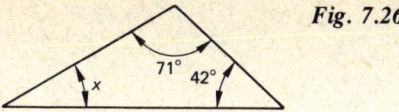

Fig. 7.26

The sum of the angles in a triangle $= 180°$
$$x + 71° + 42° = 180$$
$$x + 113° = 180°$$
$$x = 180° - 113° = 67°$$

Example 7.20 Determine the angle x in each of the triangles shown in fig. 7.27.

(a)

Fig. 7.27

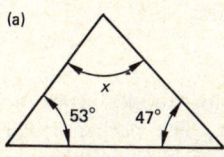

(b)

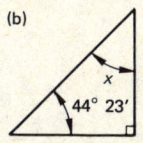

(c)

(d)

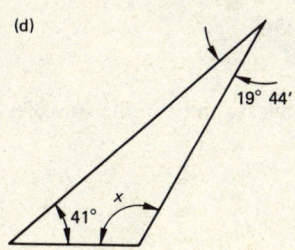

a) $x + 53° + 47° = 180°$
$\qquad x + 100° = 180°$
$x = 180° - 100° = 80°$ (*Ans.*)

b) $x + 44° 23' + 90° = 180°$
$\qquad x + 134° 23' = 180°$
$\qquad x = 180° - 134° 23' = 45° 37'$ (*Ans.*)

c) $x + 102° + 25° = 180°$
$\qquad x + 127° = 180°$
$\qquad x = 180° - 127° = 53°$ (*Ans.*)

d) $x + 19° 44' + 41° = 180°$
$\qquad x + 60° 44' = 180°$
$\qquad x = 180° - 60° 44' = 119° 16'$ (*Ans.*)

Example 7.21 Calculate the angles x, y and z on the plate template shown in fig. 7.28.

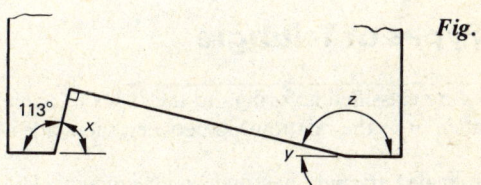

Fig. 7.28

$x + 113° = 180°$ (supplementary angles)
$x = 180° - 113° = 67°$

$x + y + 90° = 180°$ (sum of angles in a triangle)
$67° + y + 90° = 180°$
$\qquad y + 157° = 180°$
$y = 180° - 157° = 23°$

$y + z = 180°$ (supplementary angles)
$23° + z = 180°$
$z = 180° - 23° = 157°$

$\therefore$ $x = 67°$ $y = 23°$ $z = 157°$ (*Ans.*)

2   The angle z in Example 7.21 is an **exterior angle** of the triangle and could be found by using the theorem:

The exterior angle of a triangle is equal to the sum of the two opposite interior angles (fig. 7.29).

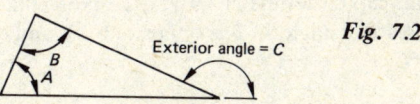

Fig. 7.29

Exterior angle $= C$

Thus considering the triangle from Example 7.21, then

$z = 67° + 90° = 157°$

Fig. 7.30

$x = 67°$

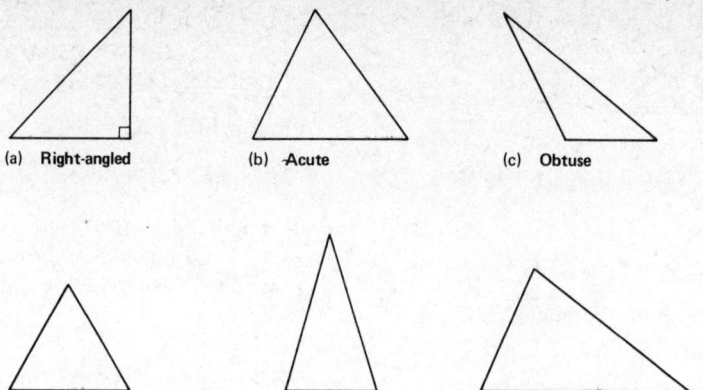

Fig. 7.31

(a) **Right-angled** (b) **Acute** (c) **Obtuse**

(d) **Equilateral** (e) **Isosceles** (f) **Scalene**

7.7 Types of Triangle

Triangles are classified according to the types of angle they contain or to the relationship between their sides.

a) **Right-angled triangle** has one angle equal to 90°. (This triangle may be solved by Pythagoras as shown in Chapter 4.)

b) **Acute-angled triangle** has *all* angles less than 90°.

c) **Obtuse-angled triangle** has one angle between 90° and 180°.

d) **Equilateral triangle** has all sides equal in length and each angle is 60°.

e) **Isosceles triangle** has two sides equal and two angles equal.

f) **Scalene triangle** has all three sides of different length.

The isosceles triangle is of particular importance in workshop calculations because many workshop shapes contain this triangle, e.g. point of a twist drill, cutting edge of a chisel, vee-thread.

Fig. 7.32 shows an isosceles triangle. The sides are denoted by the small letters a, b and c; the angles are denoted by the capital letters A, B and C. Note that side a lies opposite to angle A, b is opposite to B, and c is opposite to C.

The dotted line is a perpendicular dropped from the apex of the triangle to its base. The perpendicular bisects the side c and bisects the angle C, as shown in the triangle in fig. 7.33.

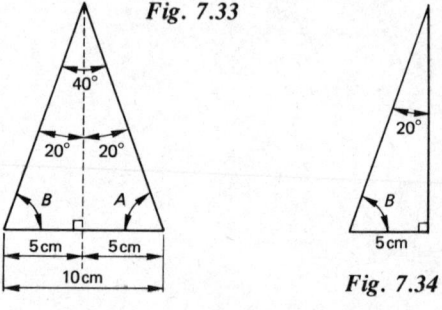

Fig. 7.33

Fig. 7.34

Consider the left-hand half of the triangle (fig. 7.34).

$B + 20° + 90° = 180°$ (sum of the angles in a triangle)

$B + 110° = 180°$

$B = 180° - 110° = 70°$

The angles A and B are equal.

$\therefore \quad A = 70°$

Example 7.22 Find the angles x and y in the isosceles triangles shown in fig. 7.35.

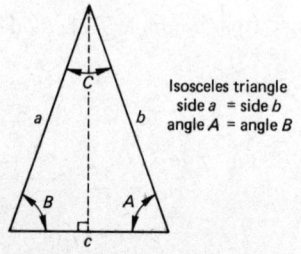

Fig. 7.32

Isosceles triangle
side a = side b
angle A = angle B

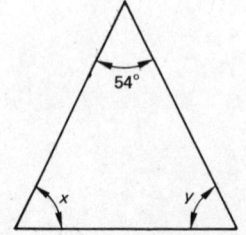

Fig. 7.35

By dropping a perpendicular from the apex we can consider half of the triangle (fig. 7.36).

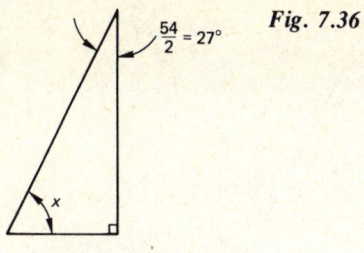

Fig. 7.36

$$\frac{54}{2} = 27°$$

$$x + 27° + 90° = 180°$$
$$x + 117° = 180°$$
$$x = 180° - 117° = 63°$$

Also $x = y$

$\therefore$ $y = 63°$

Therefore $x = 63°$ $y = 63°$ (*Ans.*)

Example 7.23   Fig. 7.37 shows the cutting edge of a chisel having a point angle of 60° and held at an angle of inclination of 40°. Calculate the value of the clearance angle and the rake angle.

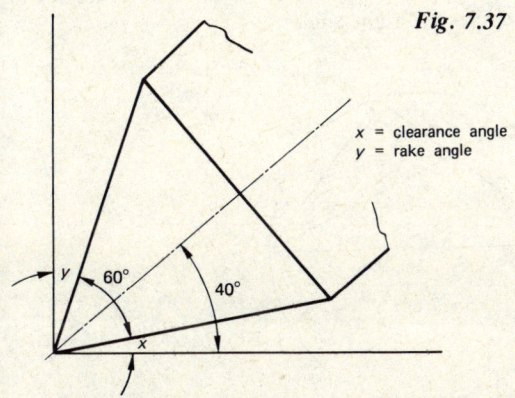

Fig. 7.37

x = clearance angle
y = rake angle

The point of the chisel is an isosceles triangle, therefore the point angle of 60° is bisected by the centre line (fig. 7.38).

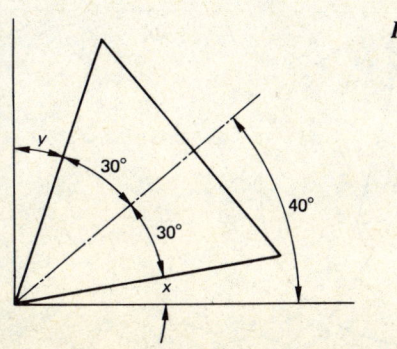

Fig. 7.38

$$x = 40° - 30°$$
$$= 10°   \text{(clearance angle)}$$
$$y = 90° - 30° - 30° - x$$
$$= 90° - 30° - 30° - 10°$$
$$= 20°   \text{(rake angle)}$$

Therefore   clearance angle $= 10°$

rake angle $= 20°$ (*Ans.*)

Example 7.24   Fig. 7.39 shows a precision steel block in the shape of an equilateral triangle. A 15° angle slip is wrung onto one side of the block. Determine the angles x and y.

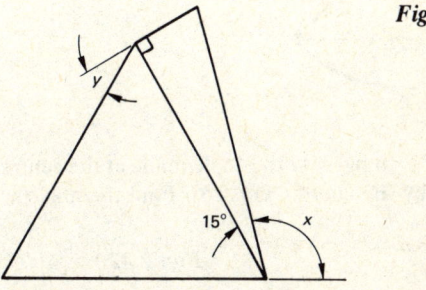

Fig. 7.39

All angles of the block are 60° (equilateral triangle).

$\therefore$ $x + 15° + 60° = 180°$
$$x + 75° = 180°$$
$$x = 180° - 75° = 105°$$

$$y + 60° + 90° = 180°$$
$$y + 150° = 180°$$
$$y = 180° - 150° = 30°$$

Therefore $x = 105°$ $y = 30°$ (*Ans.*)

7.8 Useful Angle Theorems

Fig. 7.40 shows definitions of the components of a circle referred to in this section.

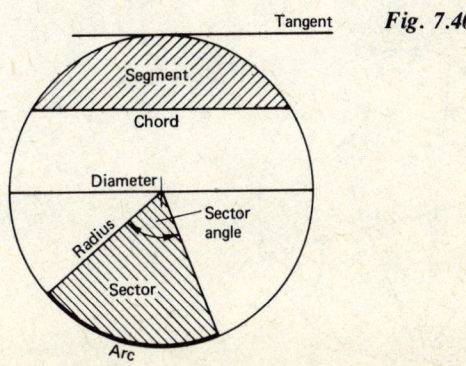

Fig. 7.40

Tangent

Segment

Chord

Diameter

Sector angle

Radius

Sector

Arc

Theorem 1 (fig. 7.41) The angle made at the centre of a circle by an arc or chord is twice the angle made at the circumference.

$$B = 2A$$

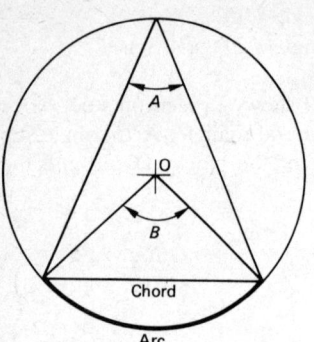

Fig. 7.41

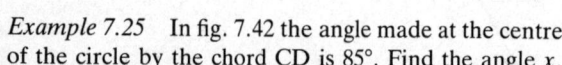

Example 7.25 In fig. 7.42 the angle made at the centre of the circle by the chord CD is 85°. Find the angle x.

Fig. 7.42

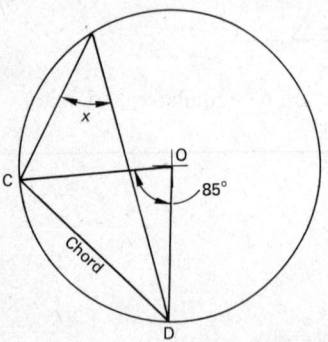

Angle at centre = 2 × angle at circumference

$$85° = 2x$$

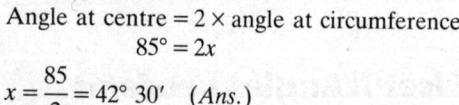

$$x = \frac{85}{2} = 42° \ 30' \quad (Ans.)$$

Theorem 2 (fig. 7.43) All angles made by an arc or chord at the circumference in the same sector of a circle are equal.

Fig. 7.43

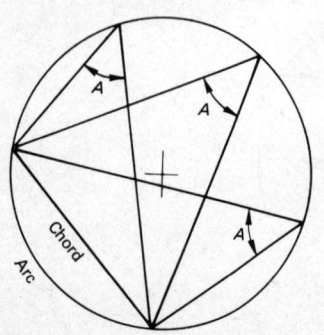

Example 7.26 Find the angles x and y in fig. 7.44.

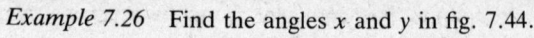

Fig. 7.44

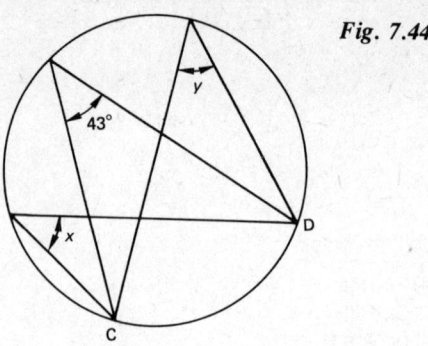

All the angles are made by the same arc CD and are therefore equal.

$$x = y = 43°$$

∴ x = 43° y = 43° (Ans.)

Theorem 3 (fig. 7.45) The angle made by the diameter at the circumference of a circle is 90°, i.e. angles in semi-circles are right-angles.

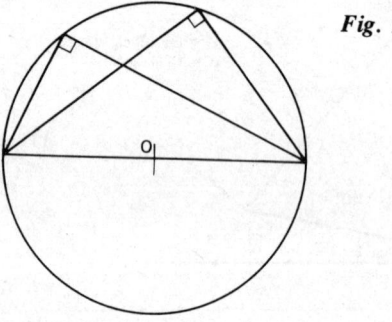

Fig. 7.45

Example 7.27 Calculate the angles x and y in fig. 7.46, where AB is the diameter of the circle.

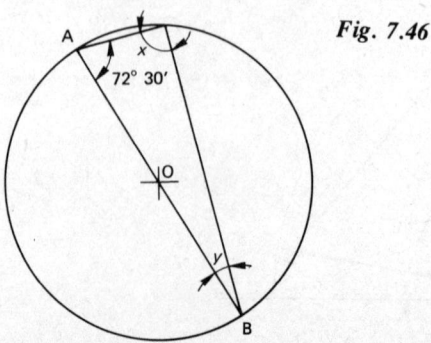

Fig. 7.46

From theorem 3, angles in semi-circles are right-angles.

$x = 90°$

$y + 72° 30' + 90° = 180°$ (sum of the angles in a triangle)

$y + 162° 30' = 180°$

$y = 180° - 162° 30' = 17° 30'$

Therefore $x = 90°$ $y = 17° 30'$ (*Ans.*)

Theorem 4 (fig. 7.47) A tangent to a circle is at right-angles to the radius at the point of contact. Two tangents drawn from a point outside the circle are of equal length.

$PM = PN$

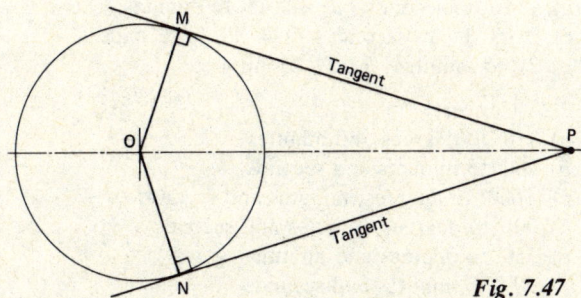

Fig. 7.47

Example 7.28 Determine the angles x and y in fig. 7.48.

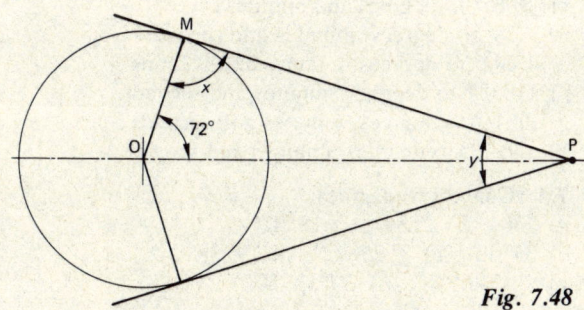

Fig. 7.48

PM is a tangent, hence x is a right-angle

$x = 90°$

$x + \tfrac{1}{2}y + 72° = 180°$ (sum of the angles in a triangle)

$90° + \tfrac{1}{2}y + 72° = 180°$

$\tfrac{1}{2}y + 162° = 180°$

$\tfrac{1}{2}y = 180° - 162° = 18°$

$y = 2 \times 18° = 36°$

Therefore $x = 90°$ $y = 36°$ (*Ans.*)

Example 7.29 Calculate the distance x in fig. 7.49.

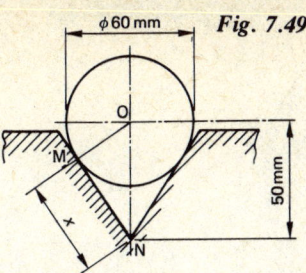

Fig. 7.49

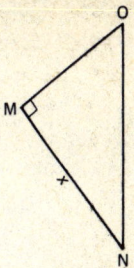

Fig. 7.50

Consider triangle OMN (fig. 7.50).

MN is a tangent, hence the angle at M is a right-angle.

OM is a radius $\quad OM = \dfrac{60}{2} = 30\,\text{mm}$

By Pythagoras:

$30^2 + x^2 = 50^2$

$900 + x^2 = 2\,500$

$x^2 = 2\,500 - 900$

$= 1\,600$

$x = \sqrt{1\,600} = 40\,\text{mm}$ (*Ans.*)

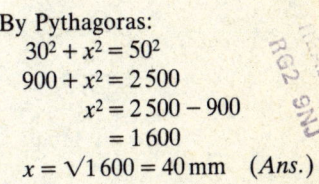

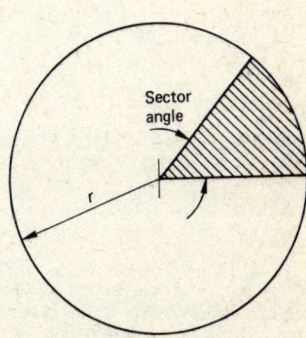

Fig. 7.51

Theorem 5 (fig. 7.51) The shaded area is a *sector* of a circle, and for any sector

$$\text{Area of a sector} = \frac{\text{sector angle}}{360} \times \text{area of circle}$$

$$= \frac{\text{sector angle}}{360} \times \pi r^2$$

Example 7.30 Calculate the area of the shaded portion of the 14 cm diameter circle shown in fig. 7.52.

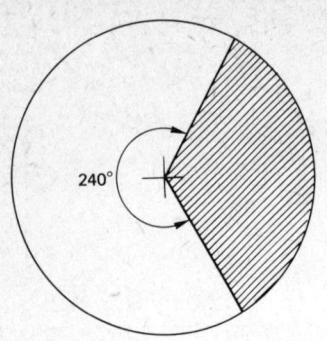

Fig. 7.52

Radius of circle $r = \dfrac{14}{2} = 7\,\text{cm}$

Sector angle $= 360° - 240° = 120°$

Area of sector $= \dfrac{\text{sector angle}}{360} \times \pi r^2$

$$= \dfrac{120}{360} \times \dfrac{22}{7} \times 7 \times 7$$

$$= 51.33\,\text{cm}^2 \quad (Ans.)$$

Theorem 6 (fig. 7.53) A *quadrilateral* is any four-sided figure. The sum of the interior angles of a quadrilateral is 360°

$A + B + C + D = 360°$

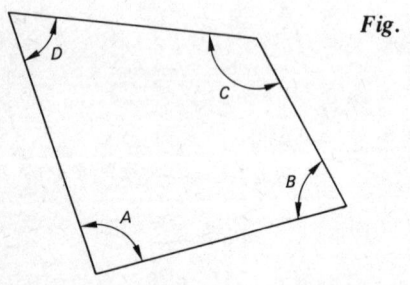

Fig. 7.53

Example 7.31 Find the angles x, y and z in fig. 7.54.

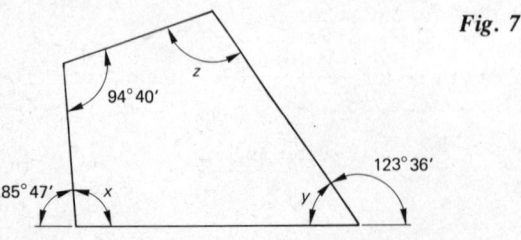

Fig. 7.54

$x + 85° \; 47' = 180°$ (supplementary angles)
$x = 180° - 85° \; 47' = 94° \; 13'$

$y + 123° \; 36' = 180°$ (supplementary angles)
$y = 180° - 123° \; 36' = 56° \; 24'$

$x + y + z + 94° \; 40' = 360°$ (angles in a quad.)
$94° \; 13' + 56° \; 24' + z + 94° \; 40' = 360°$
$z + 245° \; 17' = 360° \qquad z = 114° \; 43'$

Therefore
$\quad x = 94° \; 13' \quad y = 56° \; 24' \quad z = 114° \; 43' \quad (Ans.)$

Exercises 7

7.1 Convert
a) 3° 54′ to minutes   b) 7′ 15″ to seconds
c) 1° 10′ to seconds   d) 12° 42′ to minutes
e) 2° 09′ 14″ to seconds   f) 4° 35′ to seconds
g) $2\frac{1}{2}°$ to minutes   h) $5\frac{3}{4}°$ to minutes

7.2 Convert
a) 110′ to degrees and minutes
b) 200″ to minutes and seconds
c) 4 000″ to degrees, minutes and seconds
d) $90\frac{1}{2}'$ to degrees, minutes and seconds
e) 350′ to degrees and minutes
f) 700″ to minutes and seconds
g) 6 450″ to degrees, minutes and seconds
h) 16 000″ to degrees, minutes and seconds

7.3 Convert
a) 14.5° to degrees and minutes
b) 31.45° to degrees and minutes
c) 18.65° to degrees and minutes
d) 7.28° to degrees, minutes and seconds
e) 13.42° to degrees, minutes and seconds
f) 118.09° to degrees, minutes and seconds
g) 28.16° to degrees, minutes and seconds
h) 130.73° to degrees, minutes and seconds

7.4 Convert to degrees
a) 150′ b) 2° 36′ c) 18° 42′
d) 33° 15′ e) 8° 25′ 12″ f) 19° 38′ 24″
g) 104° 46′ 48″ h) 20° 13′ 30″

7.5 a) 32° 14′ + 51° 29′ + 7° 48′
b) 104° 33′ + 17° 54′ + 23° 07′
c) 28′ 30″ + 14′ 44″ + 54′ 08″
d) 13° 34′ 54″ + 9° 38′ 24″
e) 74° 13′ 15″ − 28° 32′ 48″
f) 35° 21′ 04″ + 16° 04′ 21″ + 49° 44′ 53″
g) Find the average of: 17° 28′, 16° 59′, 18° 09′, 17° 40′
h) Find the average: 41° 35′ 16″, 39° 28′ 46″, 40° 17′ 16″

7.6 a) Measurements of the included angle of six taper shafts are shown below:
10° 17′   10° 14′   10° 06′
9° 58′   10° 07′   9° 42′

Fig. 7.55

(a)
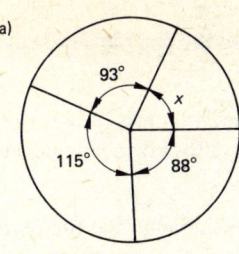

(b)
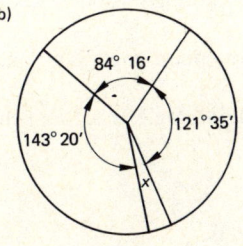

(c)
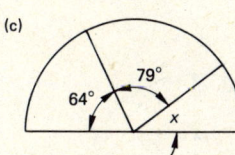

(d)
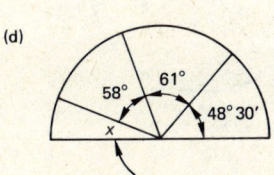

Fig. 7.56

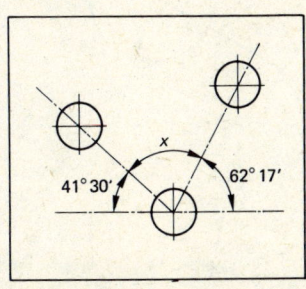

Fig. 7.57

(a)
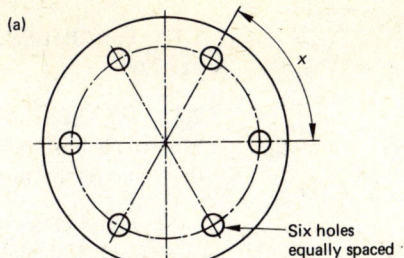

Six holes
equally spaced

(b)
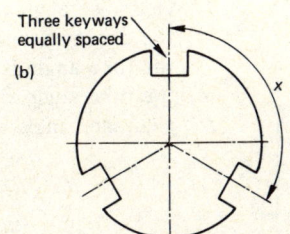

Three keyways
equally spaced

(c)
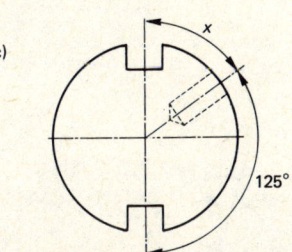

Fig. 7.58

(a)
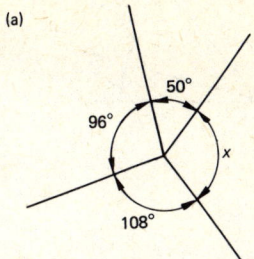

(b)
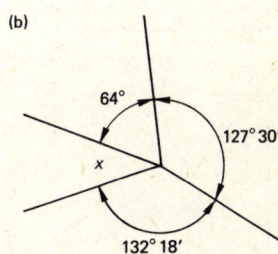

(c)
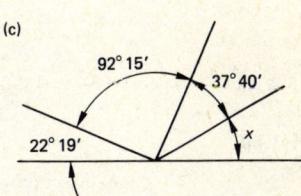

(d)
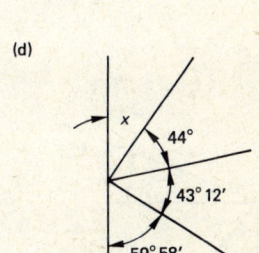

Determine the mean value of the angle.
b) Calculate the average of
41.4°, 43.6°, 40.8°, 44.2°, 39.5°

7.7 The angle x is required in each of the diagrams shown in fig. 7.55.

7.8 Calculate the angle x between the holes A and B in the drilled plate shown in fig. 7.56.

7.9 The angle x is required for each of the components shown in fig. 7.57.

7.10 Determine the angle x in each of the diagrams shown in fig. 7.58.

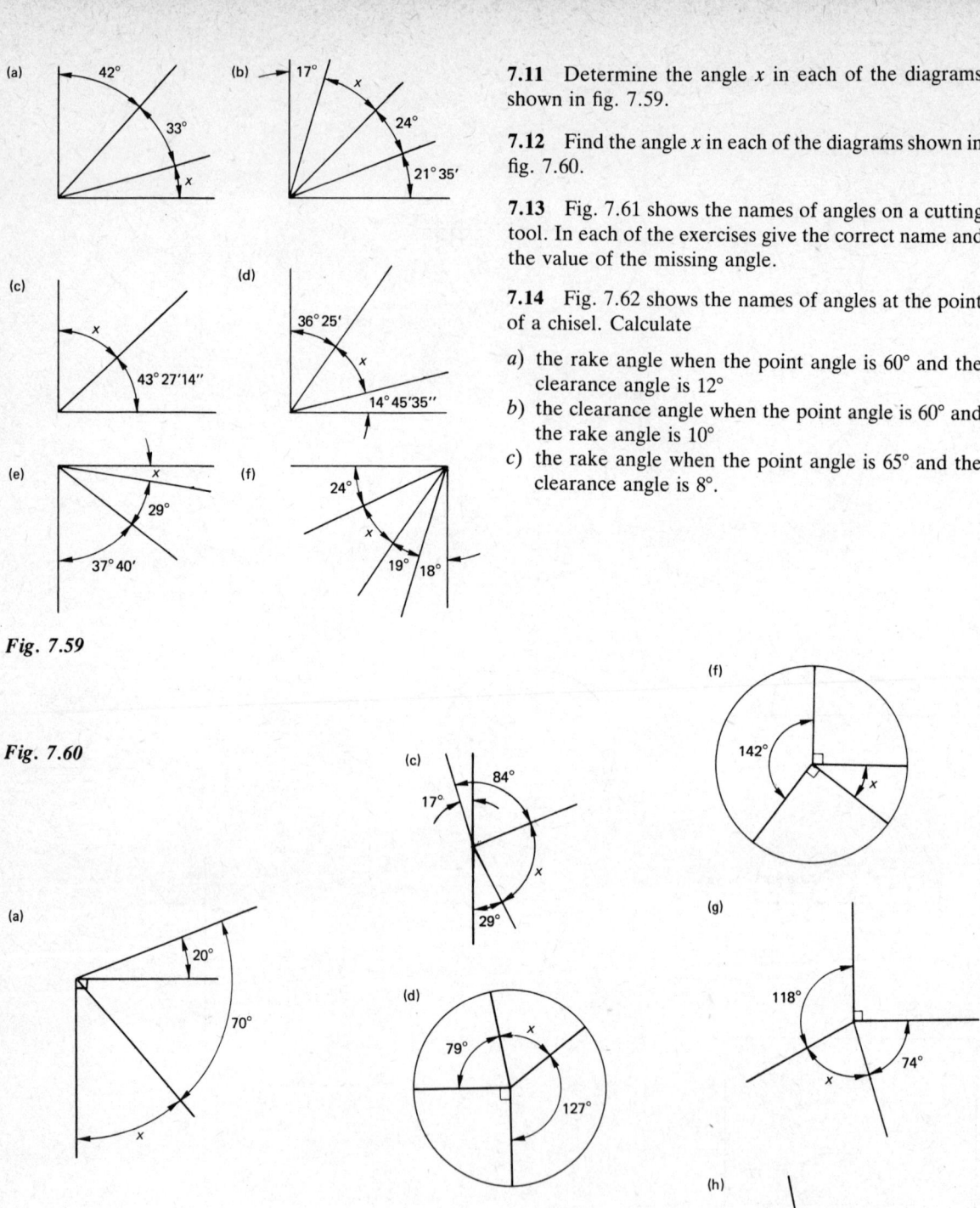

7.11 Determine the angle x in each of the diagrams shown in fig. 7.59.

7.12 Find the angle x in each of the diagrams shown in fig. 7.60.

7.13 Fig. 7.61 shows the names of angles on a cutting tool. In each of the exercises give the correct name and the value of the missing angle.

7.14 Fig. 7.62 shows the names of angles at the point of a chisel. Calculate

a) the rake angle when the point angle is 60° and the clearance angle is 12°

b) the clearance angle when the point angle is 60° and the rake angle is 10°

c) the rake angle when the point angle is 65° and the clearance angle is 8°.

Fig. 7.59

Fig. 7.60

Fig. 7.61

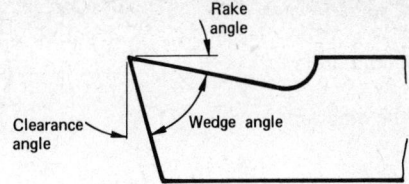

Rake angle

Clearance angle   Wedge angle

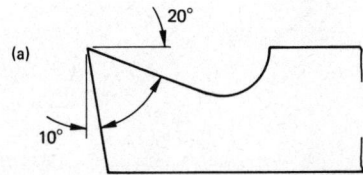

(a)

20°

10°

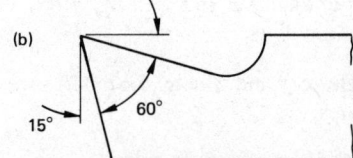

(b)

15°   60°

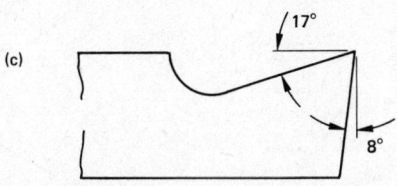

(c)

17°

8°

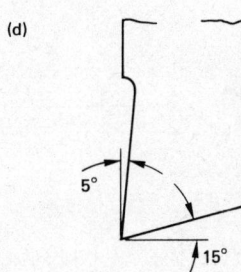

(d)

5°

15°

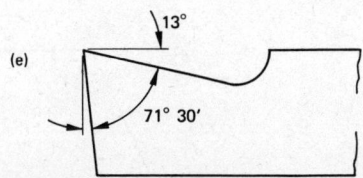

(e)

13°

71° 30'

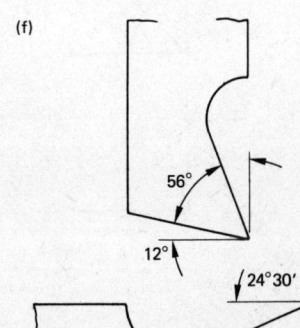

(f)

56°

12°

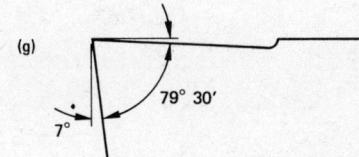

(g)

79° 30'

7°

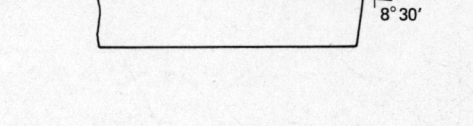

(h)

24° 30'

8° 30'

Fig. 7.62

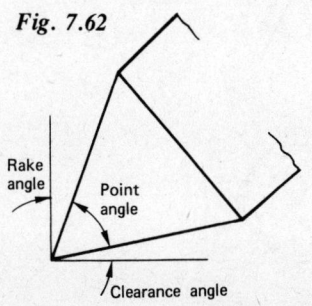

Rake angle

Point angle

Clearance angle

7.15 Place each of the following angles under its correct heading in the table below:

a) 48°   *b)* 200°   *c)* 85°   *d)* 130°
e) 175°   *f)* 60°   *g)* 300°   *h)* 100°

Acute Angle	Obtuse Angle	Reflex Angle

7.16 Give the correct name of each of the angles in fig. 7.63; take names from the following list: acute, obtuse, reflex, complementary, supplementary

7.17 Find the angles x, y and z in each of the diagrams in fig. 7.64.

7.18 Determine all the missing angles in fig. 7.65. The arrows indicate parallel lines.

7.19 Determine the angles x and y in fig. 7.66. The arrows indicate parallel lines.

7.20 Find the angles x, y and z in fig. 7.67. The arrows indicate parallel lines.

Fig. 7.65

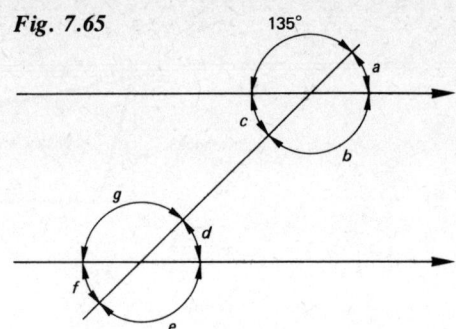

Fig. 7.63

(a)

(b)

(c)

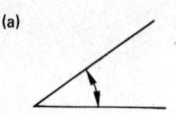
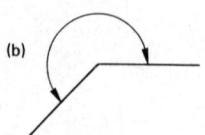
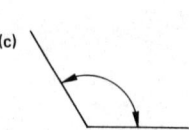

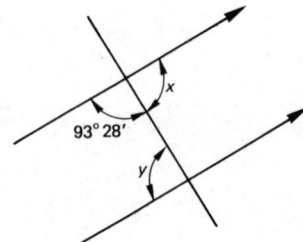

Fig. 7.66

(d)

(e)

(f)

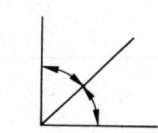
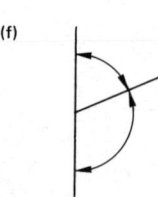

(h)

(g)

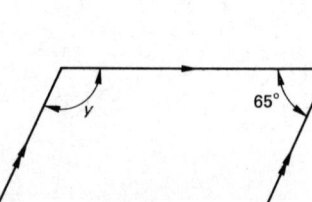

Fig. 7.67

Fig. 7.64

(a)

(b)

(c)

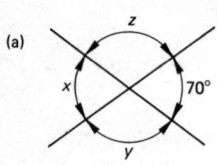
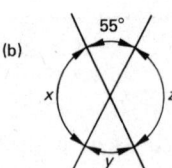
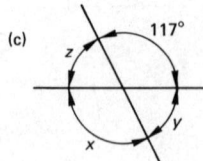

(d)

(e)

(f)

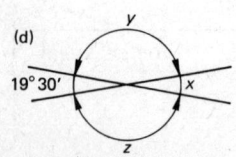
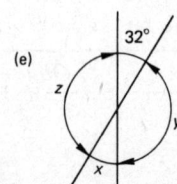
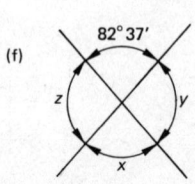

90 Workshop Numeracy

7.21 Find the angle x in each of the triangles shown in fig. 7.68.

7.22 Find the angle x in each of the isosceles triangles in fig. 7.69.

7.23 Fig. 7.70 shows the point of a chisel. If the point angle is 60° and the angle of inclination is 50°, calculate the value of the clearance angle and the rake angle.

7.24 Fig. 7.71 shows a workpiece which has been reduced in diameter in a lathe operation by the longitudinal feed of the roughing tool shown. Calculate the angles x and y on the workpiece.

Fig. 7.68

(a)
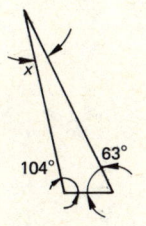

(b)
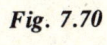

(c)
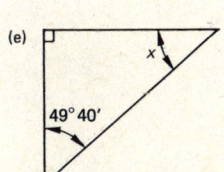

(d)
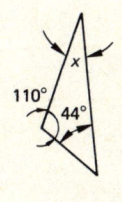

(e)
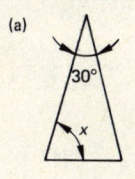

(f)
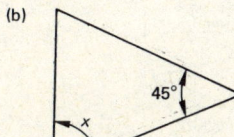

Fig. 7.70

a = rake angle
b = point angle
c = clearance angle
d = angle of inclination

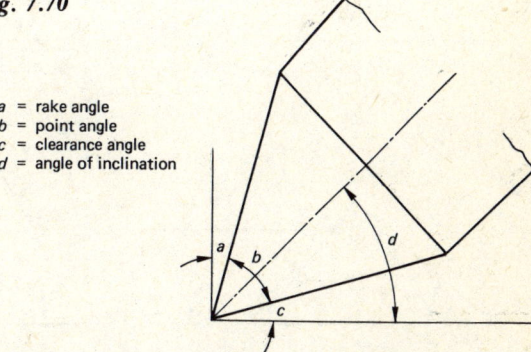

Fig. 7.69

(a)
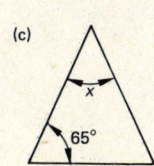

(b)
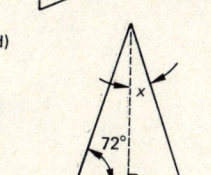

(c)
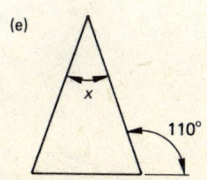

(d)

(e)

(f)
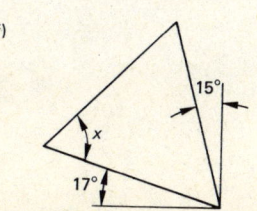

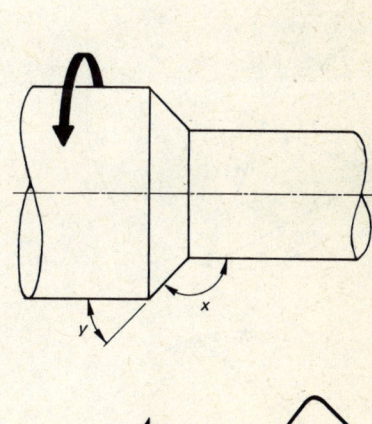

Longitudinal feed

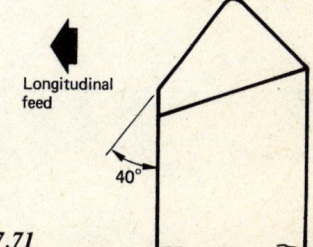

Fig. 7.71

Angles 91

7.25 Determine the angles marked x and y in each of the diagrams in fig. 7.72.

7.26 Calculate the angle x in each of the circles in fig. 7.73.

7.27 Fig. 7.74 shows two tangents to a circle drawn from the point P. Find the angles x and y.

7.28 In fig. 7.75 the lines PM and PN are tangents; calculate the distance L.

Fig. 7.72

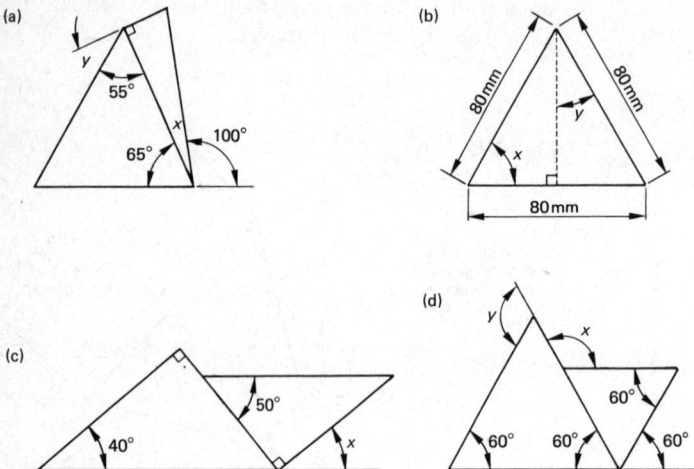

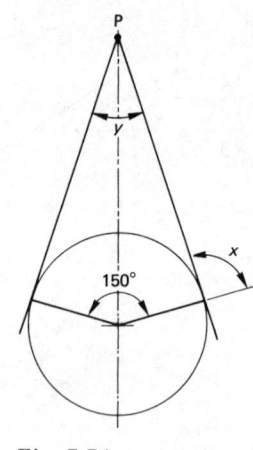

Fig. 7.74

Fig. 7.73

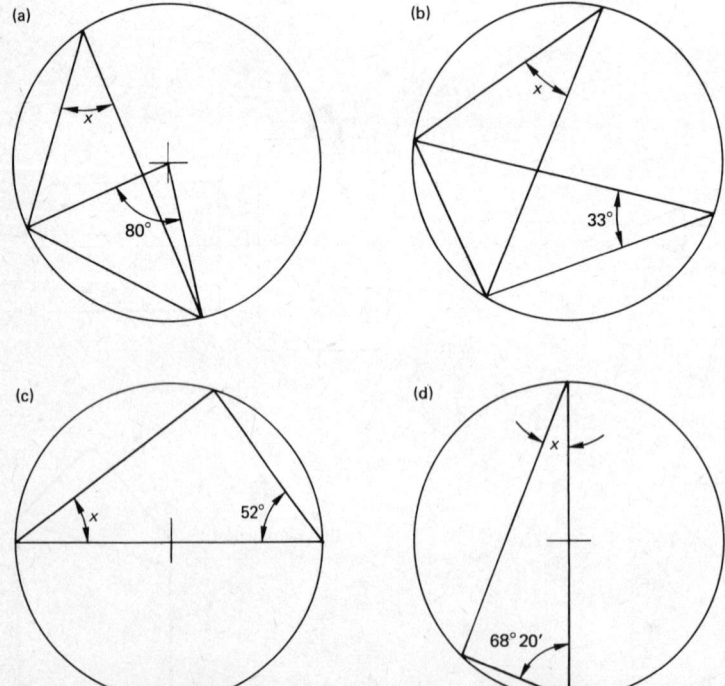

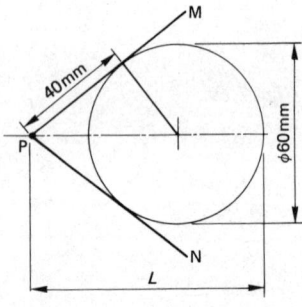

Fig. 7.75

7.29 Calculate the area of each of the following sectors:

a) 30° sector of 70 mm diameter circle.
b) 48° sector of 14 cm diameter circle.
c) 120° sector of 20 mm diameter circle.

7.30 Determine the angles marked *x*, *y* or *z* in the quadrilaterals shown in fig. 7.76.

7.31 The circular dial of an electrical instrument has 40 equally spaced divisions. Calculate the angular movement of the instrument pointer to register a difference in reading of:
a) 1 division *b)* 27 divisions

7.32 Find the number of divisions corresponding to a pointer movement of 288° on the instrument described in Exercise 7.31.

7.33 A motor shaft coupling has 5 equally spaced holes. Calculate the angle between any two adjacent holes.

7.34 Fig. 7.77 shows the spacing of the fastening holes in a motor cowling. Find the value of the angle *x*.

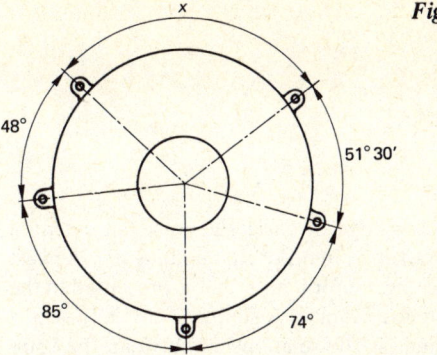

Fig. 7.77

7.35 The end ring of the rotor of an induction motor has 22 equally spaced fins to give air circulation. Calculate the angle between adjacent fins correct to the nearest minute.

Fig. 7.76

(a)

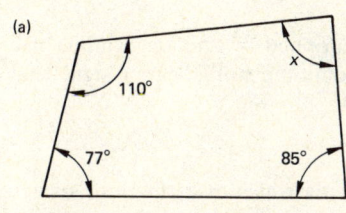

(b)

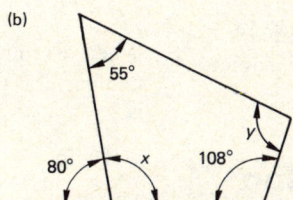

(c)

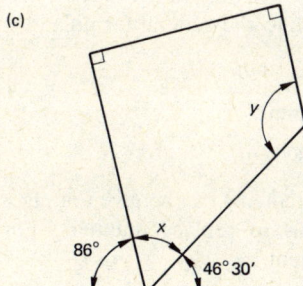

(d)

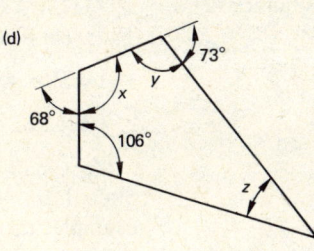

(e)

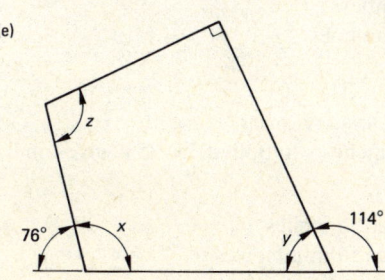

(f)

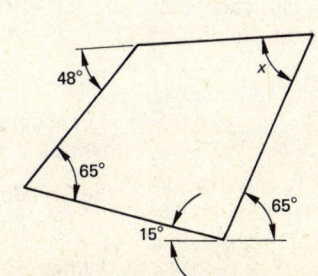

Angles 93

8 Practical Algebra

A working knowledge of basic algebra is an essential tool for the craftsman in order that he may use the vast amount of technical information that is presented in the form of algebraic formulae. A simple example of such a practical formula is the relationship between the spindle speed N of a lathe and the speed at which a material can be cut V:

$$N = \frac{1\,000\,V}{\pi d} \quad \text{where } d = \text{diameter of workpiece.}$$

The use of this formulae enables a turner to set the machine speed to give the most efficient cutting conditions for a given combination of workpiece and tool materials.

Algebra is a simple and effective tool in workshop calculations provided that the basic rules of manipulation are understood.

8.1 Symbolic Representation

In algebra, letters or other **symbols** are used in the place of numbers to allow a **general formula** to be constructed that will apply to all similar situations.

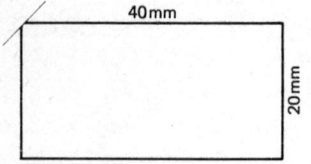

Fig. 8.1

Fig. 8.1 gives the dimensions of a particular rectangle. The area of the rectangle is given by the product of its length and breadth:

Area of rectangle = 40 mm × 20 mm
= 800 mm²

However, this numerical relationship is only true for this particular rectangle. By representing the numbers by symbols, a general relationship can be obtained which is true for *all* rectangles.

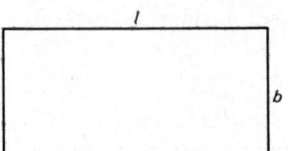

Fig. 8.2

In fig. 8.2 the length is represented by the letter l, and the breadth (width) by the letter b. If the area is represented by the letter A, then a formula can be constructed:

Area of rectangle = length × breadth

With length = l, breadth = b, area = A

$$A = l \times b$$

When a symbol is multiplied by another symbol or a number, it is usual to omit the multiplication sign. That is

$$A = lb$$

This is a general formula which gives the area of any rectangle whatever its size.

To find the area of the rectangle in fig. 8.1, we can substitute its dimensions into the general formula.

Thus, $l = 40$ mm and $b = 20$ mm

$$A = lb = 40 \times 20 = 800 \text{ mm}^2$$

∴   Area of rectangle = 800 mm²

Most mathematical statements can be given in algebraic form using symbols to replace numbers. For example, take the statement

"Two numbers multiplied together give a product."

Let a = first number
b = second number
P = product

Hence $P = ab$

For example: "inches multiplied by 25.4 gives millimetres".

Let x = dimension in inches
y = dimension in millimetres

Hence $y = 25.4x$

Example 8.1 The total cost of producing a component is made up of machining costs at £8 per hour, fitting costs at £7 per hour, and welding costs at £10 per hour. Construct an algebraic expression for the total cost of producing one component.

Let x = machining time in hours
  y = fitting time in hours
  z = welding time in hours
  C = total cost in £s of producing one component.

The cost per component of each operation
= cost per hour × time taken in hours

$\therefore$ Machining cost per component = $8 \times x$
$= 8x$
Fitting cost per component = $7 \times y$
$= 7y$
Welding cost per component = $10 \times z$
$= 10z$

Total cost per component
= machining cost + fitting cost + welding cost

$\therefore$ $C = 8x + 7y + 10z$ (*Ans.*)

Example 8.2 Use the expression derived in Example 8.1 to calculate the total cost of a component from the following information:

Machining time per component = 3 hours
  Fitting time per component = 2 hours
  Welding time per component = $\frac{1}{2}$ hour

$\therefore$ $x = 3$ $y = 2$ $z = \frac{1}{2}$

Substituting these values in the expression:

$C = 8x + 7y + 10z$

gives

$C = (8 \times 3) + (7 \times 2) + (10 \times \frac{1}{2})$
$= 24 + 14 + 5 = 43$

$\therefore$ Total cost per component = £43 (*Ans.*)

8.2 Like and Unlike Terms

Two algebraic quantities such as $5x$ and $3x$ are said to be **like terms** because they are both multiples of the same symbol x. Two quantities such as $5x$ and $3y$ are **unlike terms** because they are multiples of the different symbols x and y.

Only like terms may be added or subtracted to simplify algebraic expressions. For example

$5x + 3x = (5 + 3)x = 8x$
$5x - 3x = (5 - 3)x = 2x$

Nothing further can be done with $5x + 3y$ however.

Example 8.3 Simplify the following expressions where possible:

(i) $2y + 5y + 7y$ (ii) $6x + 4x - 3x$
(iii) $2a + 3b - 4c$ (iv) $4a + 6b + 3a - 4b$

(i) Like terms may be added.
$2y + 5y + 7y = (2 + 5 + 7)y$
$= 14y$ (*Ans.*)

(ii) Like terms may be added and subtracted.
$6x + 4x - 3x = (6 + 4 - 3)x$
$= 7x$ (*Ans.*)

(iii) Unlike terms may not be added or subtracted.
$2a + 3b - 4c$ will not simplify (*Ans.*)

(iv) Like terms may be added and subtracted.

$4a + 6b + 3a - 4b$

Rearrange the expression to bring like terms together:

$4a + 3a + 6b - 4b = (4 + 3)a + (6 - 4)b$
$= 7a + 2b$ (*Ans.*)

1 The terms x and x^2 are not like terms and cannot be added together or subtracted.

$3x + 2x^2 + 4x = 3x + 4x + 2x^2$
$= 7x + 2x^2$

Example 8.4 Simplify

$8a + 5a^2 - 2a + 4a^3 - 3a^2 + 4a$

Note that a, a^2 and a^3 are *unlike* terms.
Rearranging the expression to bring like terms together gives:

$(8a - 2a + 4a) + (5a^2 - 3a^2) + 4a^3$
$= 10a + 2a^2 + 4a^3$ (*Ans.*)

2 Like terms may be *multiplied* together using **indices** (or *powers*):

$a \times a$ $= a^2$
$a \times a \times a$ $= a^3$
$a \times a \times a \times a = a^4$

In a^4, the 4 is the index or power. So a^4 is "a to the power 4".

3 In terms such as $3a$, the number 3, which multiplies a, is called the **coefficient** of a. The coefficients of terms must also be multiplied together, thus

$3a \times 2a = (3 \times 2)(a \times a)$
$= 6a^2$

$2a \times 3a \times 4a = (2 \times 3 \times 4)(a \times a \times a)$
$= 24a^3$

So the coefficients are multiplied together as one step,

and the symbols are multiplied together as another separate step.

The coefficients of unlike terms also multiply together, thus

$$2a \times 3b = (2 \times 3)(a \times b)$$
$$= 6ab$$

4 Brackets contained in expressions should be expanded by multiplication before simplifying the expression. For example

$$3(a + b) + 2a = (3a + 3b) + 2a$$
$$= 5a + 3b$$

$$4(3x - y) + 7(x + 2y) = (12x - 4y) + (7x + 14y)$$
$$= 12x + 7x - 4y + 14y$$
$$= 19x + 10y$$

Example 8.5 Simplify:

(i) $7a \times 2a$ (iv) $7a \times 4b \times 2c$
(ii) $3y \times 2y^2$ (v) $5(3a + 2b) + 2(a + c)$
(iii) $2b \times 3b \times 5b^2$ (vi) $6(3x + 2) - 2(x + 1)$

(i) $7a \times 2a = (7 \times 2)(a \times a)$
$$= 14a^2 \quad (Ans.)$$

(ii) $3y \times 2^2 = (3 \times 2)(y \times y^2)$
$$= 6y^3 \quad (Ans.)$$

(iii) $2b \times 3b \times 5b^2 = (2 \times 3 \times 5)(b \times b \times b^2)$
$$= 30b^4 \quad (Ans.)$$

(iv) $7a \times 4b \times 2c = (7 \times 4 \times 2)(a \times b \times c)$
$$= 56abc \quad (Ans.)$$

(v) $5(3a + 2b) + 2(a + c) = (15a + 10b) + (2a + 2c)$
$$= 17a + 10b + 2c \quad (Ans.)$$

(vi) $6(3x + 2) - 2(x + 1) = (18x + 12) - (2x + 2)$
$$= 18x + 12 - 2x - 2$$
$$= 16x + 10 \quad (Ans.)$$

5 Algebraic expressions involving division by like terms may be simplified by cancelling.

$$\frac{ab}{a} = \frac{\cancel{a} \times b}{\cancel{a}} = b$$

$$\frac{8ab}{4a} = \frac{\overset{2}{\cancel{8}} \times \cancel{a} \times b}{\cancel{4} \times \cancel{a}} = 2b$$

$$\frac{12a^2}{4a} = \frac{\overset{3}{\cancel{12}} \times \cancel{a} \times a}{\cancel{4}\cancel{a}} = 3a$$

Example 8.6 Simplify:

(i) $\dfrac{20xy}{4y}$ (ii) $\dfrac{24m^2}{3m}$

(iii) $\dfrac{15a^3}{3a^2}$ (iv) $\dfrac{12b^3c^2}{3bc}$

(v) $\dfrac{8(a^2 + a)}{4a}$ (vi) $\dfrac{4(3x^2 + 6x)}{3x}$

(i) $\dfrac{20xy}{4y} = \dfrac{\overset{5}{\cancel{20}} \times x \times \cancel{y}}{\cancel{4} \times \cancel{y}} = 5x \quad (Ans.)$

(ii) $\dfrac{24m^2}{3m} = \dfrac{\overset{8}{\cancel{24}} \times \cancel{m} \times m}{\cancel{3} \times \cancel{m}} = 8m \quad (Ans.)$

(iii) $\dfrac{15a^3}{3a^2} = \dfrac{\overset{5}{\cancel{15}} \times \cancel{a} \times \cancel{a} \times a}{\cancel{3} \times \cancel{a} \times \cancel{a}} = 5a \quad (Ans.)$

(iv) $\dfrac{12b^3c^2}{3bc} = \dfrac{\overset{4}{\cancel{12}} \times \cancel{b} \times b \times b \times \cancel{c} \times c}{\cancel{3} \times \cancel{b} \times \cancel{c}}$
$$= 4 \times b \times b \times c = 4b^2c \quad (Ans.)$$

(v) $\dfrac{8(a^2 + a)}{4a} = \dfrac{8a^2 + 8a}{4a} = \dfrac{\overset{2}{\cancel{8}} \times \cancel{a} \times a}{\cancel{4}\cancel{a}} + \dfrac{\overset{2}{\cancel{8}}\cancel{a}}{\cancel{4}\cancel{a}}$
$$= 2 \times a + 2$$
$$= 2a + 2 \quad (Ans.)$$

(vi) $\dfrac{4(3x^2 + 6x)}{3x} = \dfrac{12x^2 + 24x}{3x}$
$$= \dfrac{\overset{4}{\cancel{12}} \times \cancel{x} \times x}{\cancel{3}\cancel{x}} + \dfrac{\overset{8}{\cancel{24}} \times \cancel{x}}{\cancel{3}\cancel{x}}$$
$$= 4 \times x + 8$$
$$= 4x + 8 \quad (Ans.)$$

8.3 Substitution in Algebraic Expressions

In the same way that symbols can take the place of numbers to give algebraic expressions, numbers can be substituted back in the expression in order to **evaluate** the expression:

For example, evaluate $9x + 3y$, when $x = 2$ and $y = 4$.

By substituting 2 for x and 3 for y the expression:

$$9x + 3y \text{ becomes } (9 \times 2) + (3 \times 4)$$
$$= 18 + 12 = 30$$

Example 8.7 Determine the value of $9a + 3b - 5c$, when $a = 4$, $b = 2$ and $c = 3$.

$$9a + 3b - 5c = (9 \times 4) + (3 \times 2) - (5 \times 3)$$
$$= 36 + 6 - 15$$
$$= 42 - 15$$
$$= 27 \quad (Ans.)$$

Example 8.8 Evaluate

$\dfrac{3x + y}{z}$, when $x = 6$, $y = 2$ and $z = 4$.

$$\frac{3x + y}{z} = \frac{(3 \times 6) + 2}{4} = \frac{18 + 2}{4} = \frac{20}{4} = 5 \quad (Ans.)$$

Example 8.9 Find the value of $2x^2 + 5x - y^2$ when $x = 3$ and $y = 5$.

$$2x^2 + 5x - y^2 = (2 \times 3 \times 3) + (5 \times 3) - (5 \times 5)$$
$$= 18 + 15 - 25$$
$$= 33 - 25 = 8 \quad (Ans.)$$

Example 8.10 Given that $a = 3$, $b = 9$ and $c = 2$, find the value of

(i) $2a^2 + 4b - 3c$

(ii) $a^2 + b^2 + c^2$

(iii) $\dfrac{14a - 3b}{c^2}$

(i) $2a^2 + 4b - 3c = (2 \times 3 \times 3) + (4 \times 9) - (3 \times 2)$
$$= 18 + 36 - 6 = 48 \quad (Ans.)$$

(ii) $a^2 + b^2 + c^2 = (3 \times 3) + (9 \times 9) + (2 \times 2)$
$$= 9 + 81 + 4 = 94 \quad (Ans.)$$

(iii) $\dfrac{14a - 3b}{c^2} = \dfrac{(14 \times 3) - (3 \times 9)}{(2 \times 2)}$

$$= \frac{42 - 27}{4} = \frac{15}{4} = 3.75 \quad (Ans.)$$

Example 8.11 Determine the value of P when

$$P = \frac{4K + S}{T}, \quad K = 1.8, \ S = 3.2 \text{ and } T = 0.5.$$

$$P = \frac{4K + S}{T}$$

$$= \frac{(4 \times 1.8) + 3.2}{0.5}$$

$$= \frac{7.2 + 3.2}{0.5} = \frac{10.4}{0.5} = 20.8 \quad (Ans.)$$

Some problems require the **substitution of negative values** in the expression, and the rules for the multiplication of *directed numbers* must be observed.

When multiplying *like* signs the product is positive:

$+ \times + = +$ (plus times plus is plus)
$- \times - = +$ (minus times minus is plus)

and when multiplying *unlike* signs the product is negative:

$+ \times - = -$ (plus times minus is minus)
$- \times + = -$ (minus times plus is minus)

Example 8.12 Given that $a = 2$, $b = -1$ and $c = 4$, evaluate

(i) $4a = 3b + c$

(ii) $3a^2 + b^2 + c^2$

(iii) $\dfrac{2a + c}{b}$

(i) $4a + 3b + c = (4 \times 2) + (3 \times -1) + 4$
$$= 8 + (-3) + 4$$
$$= 8 - 3 + 4$$
$$= 9 \quad (Ans.)$$

(ii) $3a^2 + b^2 + c^2 = (3 \times 2 \times 2) + (-1 \times -1) + (4 \times 4)$
$$= 12 + (+1) + 16$$
$$= 12 + 1 + 16$$
$$= 29 \quad (Ans.)$$

(iii) $\dfrac{2a + c}{b} = \dfrac{(2 \times 2) + 4}{-1}$

$$= \frac{4 + 4}{-1} = \frac{8}{-1} = -8 \quad (Ans.)$$

8.4 Simple Equations and Transposition

An **equation** is simply a statement that one quantity equals another. The following statements are all equations:

1 inch = 25.4 mm
1 km = 1 000 m
diameter = 2 × radius

In algebra, the equation usually contains an unknown quantity which is represented by a symbol such as x. For example

$2x = 6$

It is required to **solve** the equation in order to find the value of x. In this example, it is obvious that x must have the value of 3 in order that $2x$ can equal 6.

Hence, the **solution** of the equation $2x = 6$ is $x = 3$.

The types of equation to be solved at this stage of studies are called *Simple Equations* because they contain only *one* unknown and do not contain any powers of the unknown such as x^2, x^3, etc.

To solve a simple equation, the two sides may be manipulated or transposed to obtain the unknown quantity by itself on one side of the equation. This **transposition** must be done in accordance with certain rules so that the equation will *balance* at all stages, i.e. the left-hand side (L.H.S.) will always equal the right-hand side (R.H.S.).

$$\text{L.H.S.} = \text{R.H.S.}$$

The rules of transposition are as follows:

Rule 1 If a quantity is connected to one side of the equation by a + or − sign, it may move to the other side by *changing its sign*, so that + becomes −, and − becomes +.

L.H.S	=	R.H.S
+	⟷	−
−	⟷	+

Rule 2 If a quantity multiplies or divides on one side of the equation, it may move to the other side by *changing its operation*, so that × becomes ÷, and ÷ becomes ×.

L.H.S.	=	R.H.S.
×	⟷	÷
÷	⟷	×

The whole purpose of using these rules is *to isolate the unknown on one side of the equation.* Once this has been done the equation has been solved.

Example 8.13 Solve $x + 4 = 9$.

To obtain the unknown x by itself, we must move the value +4. Using Rule 1, the value +4 moves from L.H.S.→R.H.S. and changes its sign from + to −. Thus

$$x = 9 - 4$$
$$x = 5 \quad (Ans.)$$

The answer can be checked by substituting the value $x = 5$ in the original equation, which must then balance.

$$x + 4 = 9$$
$$5 + 4 = 9$$
$$9 = 9$$

so the solution $x = 5$ is correct.

Example 8.14 Solve $x - 6 = 2$.
Using Rule 1, the value −6 moves from L.H.S.→R.H.S. and changes its sign from − to +.

$$x = 2 + 6$$
$$x = 8 \quad (Ans.)$$

Example 8.15 Solve $9x = 36$.
Using Rule 2, the value 9 which is multiplying on the L.H.S. moves to the R.H.S. and changes its operation to division.

$$x = \frac{36}{9}$$
$$x = 4 \quad (Ans.)$$

Example 8.16 Solve $\frac{x}{5} = 3$.

Using Rule 2, the value 5 which is dividing on the L.H.S. moves to the R.H.S. where it multiplies.

$$x = 3 \times 5$$
$$x = 15 \quad (Ans.)$$

1 Many solutions will require the use of both rules of transposition in order to isolate the unknown on one side of the equation.

Example 8.17 Solve $3x + 5 = 32$.

Using Rule 1	$3x = 32 - 5$
	$3x = 27$
Using Rule 2	$x = \dfrac{27}{3}$
	$x = 9 \quad (Ans.)$

Example 8.18 Solve $\frac{x - 3}{6} = 7$.

Using Rule 2	$x - 3 = 7 \times 6$
	$x - 3 = 42$
Using Rule 1	$x = 42 + 3$
	$x = 45 \quad (Ans.)$

Example 8.19 Solve $\frac{2y + 8}{4} = 7$.

Using Rule 2	$2y + 8 = 7 \times 4$
	$2y + 8 = 28$
Using Rule 1	$2y = 28 - 8$
	$2y = 20$
Using Rule 2	$y = \dfrac{20}{2}$
	$y = 10 \quad (Ans.)$

2 In some equations, the unknown may appear on both sides of the equation, and it will be necessary to use the rules of transposition to collect all expressions containing the unknown to one side.

Example 8.20   Solve $3m + 2 = m + 12$.

Using Rule 1 $3m = m + 12 - 2$
$3m = m + 10$

Collect values of m (Rule 1) $3m - m = 10$
$2m = 10$

Using Rule 2 $m = \dfrac{10}{2}$
$m = 5$ (*Ans.*)

Checking the answer by substituting $m = 5$ in the equation:

$$(3 \times 5) + 2 = 5 + 12$$
$$15 + 2 = 17$$
$$17 = 17$$

Example 8.21   Solve $\frac{4}{5}k - 8 = \frac{1}{3}k + 4$.

(Rule 1) $\frac{4}{5}k = \frac{1}{3}k + 4 + 8$
$\frac{4}{5}k = \frac{1}{3}k + 12$

(Rule 1) $\frac{4}{5}k - \frac{1}{3}k = 12$
$\frac{3}{5}k = 12$

(Rule 2) $k = 12 \times \dfrac{5}{3} = \dfrac{60}{3} = 20$ (*Ans.*)

3 When brackets appear in equations, they must be removed first in accordance with the rules of directed numbers on page 97. (Multiplying like signs gives positive; multiplying unlike signs gives negative.)

Example 8.22   Solve $3(x - 4) = 9$.

$3x - 12 = 9$
$3x = 9 + 12$
$3x = 21$
$x = \dfrac{21}{3} = 7$ (*Ans.*)

Example 8.23   Solve $3(w + 4) - 5(w - 1) = 19$.

$3w + 12 - 5w + 5 = 19$
$3w - 5w = 19 - 12 - 5$
$-2w = 2$
$-w = 1$

Multiply both sides by -1 $w = -1$ (*Ans.*)

Example 8.24   Solve
$6(2x - 2.3) = 8(x + 0.7) - 12.2$

$12x - 13.8 = 8x + 5.6 - 12.2$

$12x - 8x = 5.6 - 12.2 + 13.8$
$4x = 7.2$
$x = \dfrac{7.2}{4} = 1.8$ (*Ans.*)

8.5 Transposition of Formulae

A *formula* is simply an equation and therefore the same rules of transposition may be used. In Section 8.1 a formula was derived for the area of a rectangle:

$$A = lb$$

where A = area, l = length, b = breadth.

This formulae is used to find the area when both dimensions are known. By transposition, the formula may be rearranged to give the length when the area and the breadth are known:

$$A = lb$$
$$\frac{A}{b} = l$$

which can be written $l = \dfrac{A}{b}$

Hence, this is a formula for the length of the rectangle.

In the same way the formula can be transposed to give the breadth:

$$A = lb$$
$$\frac{A}{l} = b$$
$$b = \frac{A}{l}$$

Hence, this is a formula for the breadth of the rectangle.

In a formula such as

$$N = \frac{1\,000V}{\pi d}$$

we say that N is the **subject** of the formula because it is the quantity that the formula is arranged to find. Using the rules of transposition, the same formula can be rearranged to make any of the other terms appear as the subject.

$$N = \frac{1\,000V}{\pi d}$$

To make V the subject:

$$\pi dN = 1\,000V$$
$$\frac{\pi dN}{1\,000} = V$$

$$V = \frac{\pi dN}{1\,000}$$

To make d the subject:

$$\pi dN = 1\,000V$$

$$d = \frac{1\,000V}{\pi N}$$

Example 8.25 The volume of a cylinder is given by the formula:

$$V = \pi r^2 l$$

where V = volume in mm³, r = radius in mm, l = length in mm.

Make l the subject of the formula and hence calculate the length of a cylinder of radius 10 mm and volume 13 200 mm³.

$$V = \pi r^2 l$$

$$\frac{V}{\pi r^2} = l$$

$$l = \frac{V}{\pi r^2} \quad (Ans.)$$

When $r = 10$ and $V = 13\,200$,

$$l = \frac{13\,200}{\pi \times 10^2} = \frac{13\,200}{\frac{22}{7} \times 10 \times 10} = 42\,\text{mm} \quad (Ans.)$$

Example 8.26 In each of the following make x the subject of the formula:

(i) $y = ax + b$
(ii) $k = a + b(x - 1)$

(iii) $C = \dfrac{x}{R + r}$

(iv) $V = \dfrac{\pi d^2 x}{4}$

(i) $y = ax + b$
       $y - b = ax$
       $\dfrac{y - b}{a} = x$

         $x = \dfrac{y - b}{a} \quad (Ans.)$

(ii) $k = a + b(x - 1)$
        $k - a = b(x - 1)$
        $k - a = bx - b$
      $k - a + b = bx$
      $\dfrac{k - a + b}{b} = x$

        $x = \dfrac{k - a + b}{b} \quad (Ans.)$

(iii) $C = \dfrac{x}{R + r}$

The terms $R + r$ must be moved together in a bracket:

$$C(R + r) = x$$
$$x = C(R + r) \quad (Ans.)$$

(iv) $V = \dfrac{\pi d^2 x}{4}$

     $4V = \pi d^2 x$

     $\dfrac{4V}{\pi d^2} = x$

     $x = \dfrac{4V}{\pi d^2} \quad (Ans.)$

Example 8.27 Given that $E = IR$, transpose the formula to give:

a) R in terms of the other two quantities
b) I in terms of the other two quantities.

a) $E = IR$
    $\dfrac{E}{I} = R$

    $R = \dfrac{E}{I} \quad (Ans.)$

b) $E = IR$
    $\dfrac{E}{R} = I$

    $I = \dfrac{E}{R} \quad (Ans.)$

Example 8.28 Given that $V = E + IR$, transpose the formula to give:

a) E b) I c) R

a) $V = E + IR$
    $V - IR = E$
      $E = V - IR \quad (Ans.)$

b) $V = E + IR$
    $V - E = IR$
      $IR = V - E$
      $I = \dfrac{V - E}{R} \quad (Ans.)$

c) $V = E + IR$
    $V - E = IR$
      $IR = V - E$
      $R = \dfrac{V - E}{I} \quad (Ans.)$

Example 8.29 Given that $P = I^2R$ transpose the formula to give a) I b) R

a) $P = I^2R$

$$\frac{P}{R} = I^2$$

$$I^2 = \frac{P}{R}$$

$$I = \sqrt{\frac{P}{R}} \quad \text{or} \quad I = \frac{\sqrt{P}}{\sqrt{R}} \quad (Ans.)$$

b) $P = I^2R$

$$\frac{P}{I^2} = R$$

$$R = \frac{P}{I^2} \quad (Ans.)$$

8.6 Construction of Formulae

Few problems in the workshop are entirely new and the craftsman is able to draw on his experience of similar situations in most cases. In this way he applies his accumulated general knowledge of a problem area to a specific problem he wishes to solve. Formulae are a means of expressing in a general way the relationships acting in common situations so that they may be used in solving future problems. This avoids the time-wasting approach of treating every problem as completely new.

A **formulae** is an equation which is constructed from the known data of a situation and expressed in a form which is convenient to use in all similar situations.

Fig. 8.3 shows a rectangular instrument panel having a punched hole to accept an instrument dial. A general formula can be constructed to give the area of the panel face.

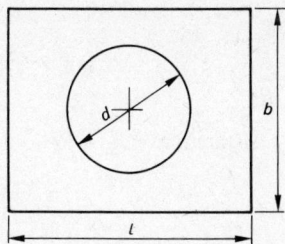

Fig. 8.3

Let l = length of panel, mm
b = breadth of panel, mm
d = diameter of hole, mm
A = area of panel, mm².

It is important to include the units of each dimension so that the formula will give an answer in the correct units of area. A general statement can be made showing the elements that make up the area of the panel:

Area of panel = area of rectangle − area of hole

These elements can now be stated from known data.

Area of rectangle = length × breadth
$$= lb$$

Area of hole = $\pi \times$ radius²

and since radius = $\dfrac{\text{diameter}}{2}$, then

$$\text{Area of hole} = \pi \times \left(\frac{\text{diameter}}{2}\right)^2$$

$$= \pi \times \left(\frac{d}{2}\right)^2$$

$$= \pi \times \frac{d}{2} \times \frac{d}{2}$$

$$= \frac{\pi d^2}{4}$$

Area of panel = area of rectangle − area of hole

$$A = lb - \frac{\pi d^2}{4} \quad (Ans.)$$

This is a general formula which can be used to find the area of *any* sized panel of the *same* shape. Whenever the formula is stated, the symbols must be defined and their units given.

Example 8.30 The area of a rectangular instrument panel having a circular hole is given by the formula:

$$A = lb - \frac{\pi d^2}{4}$$

where l = length of panel, mm
b = breadth of panel, mm
d = diameter of hole, mm
A = area of panel, mm².

Calculate the area of an instrument panel 100 mm by 60 mm having a hole of 28 mm diameter.

First list the known elements: $l = 100$ mm, $b = 60$ mm, $d = 28$ mm.

$$A = lb - \frac{\pi d^2}{4}$$

Substituting in formula

$$A = 100 \times 60 - \frac{22}{7} \times \frac{28 \times 28}{4}$$

$$= 6\,000 - 22 \times 28$$

$$= 6\,000 - 616 = 5\,384$$

Area of panel = $5\,384$ mm² (Ans.)

Example 8.31 Fig. 8.4 shows the cutting angles at the point of a chisel. Construct a formula to give the rake angle in terms of the other two angles.

Fig. 8.4

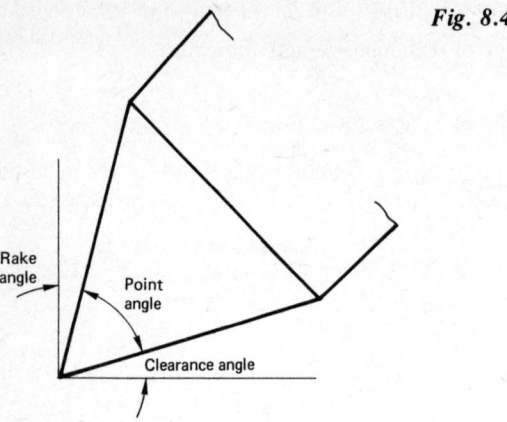

Let a = rake angle, degrees
       b = point angle, degrees
       c = clearance angle, degrees

Rake angle + point angle + clearance angle = 90°

$a + b + c = 90°$
$\qquad a = 90° - b - c$ (*Ans.*)

Example 8.32 Use the formula constructed in Example 8.31 to determine the clearance angle of a chisel when the point angle is 60° and the rake angle is 20°.

$a = 20°$ $b = 60°$

$a = 90° - b - c$
Transposing,
$c = 90° - a - b$
$\quad = 90° - 20° - 60° = 90° - 80° = 10°$

Clearance angle = 10°   (*Ans.*)

Example 8.33 a) Construct a formula to give the volume of metal removed when machining the tee-slot shown in fig. 8.5 in a cast-iron block l mm long.

Fig. 8.5

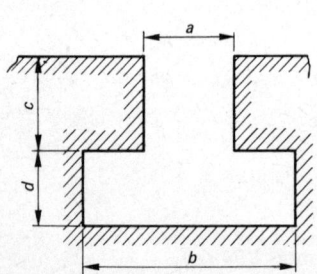

b) Use the formula to calculate the volume of metal removed when machining a tee-slot of dimensions, $a = 16$ mm, $b = 30$ mm, $c = 15$ mm and $d = 12$ mm in a block of metal 200 mm long.

a)   Let V = volume of metal removed, mm³

V = cross-sectional area (C.S.A.) of slot × length
C.S.A. of slot = area of rectangle $a \times c$
$\qquad\qquad\qquad$ + area of rectangle $b \times d$
$\qquad\qquad = ac + bd$
$\therefore \quad V = (ac + bd) \times l$
$\qquad\quad = l(ac + bd)$ (*Ans.*)

b) $a = 16$ mm $b = 30$ mm $c = 15$ mm
$d = 12$ mm $l = 200$ mm

$V = l[ac + bd]$
$\quad = 200[(16 \times 15) + (30 \times 12)]$
$\quad = 200[240 + 360]$
$\quad = 200 \times 600 = 120\,000$

Volume of metal removed = 120 000 mm³   (*Ans.*)

Example 8.34 a) Construct a formula to find the current I flowing in the circuit shown in fig. 8.6.
b) Use the formula to find the current flowing when $R_1 = 18$ ohms, $R_2 = 30$ ohms, $E = 24$ volts.

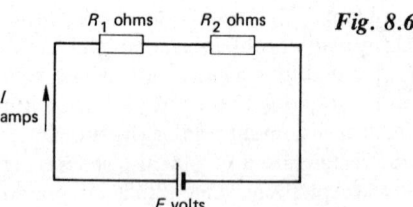

Fig. 8.6

a)   Using Ohms Law $E = IR$

Transposing $\qquad I = \dfrac{E}{R}$

In a series circuit, the total resistance $R = R_1 + R_2$

$\therefore \quad I = \dfrac{E}{R_1 + R_2}$ (*Ans.*)

b) $I = \dfrac{E}{R_1 + R_2}$

$\quad = \dfrac{24}{18 + 30}$

$\quad = \dfrac{24}{48} = \dfrac{1}{2} = 0.5$ amp   (*Ans.*)

8.7 Formula for Cutting Speed

1 In the lathe operation shown in fig. 8.7, the workpiece of diameter d mm is rotated at a spindle speed of N revolutions per minute. The speed at which the workpiece surface is passing the tool point is called the **cutting speed** of the operation and is given as V metres per minute.

Fig. 8.7

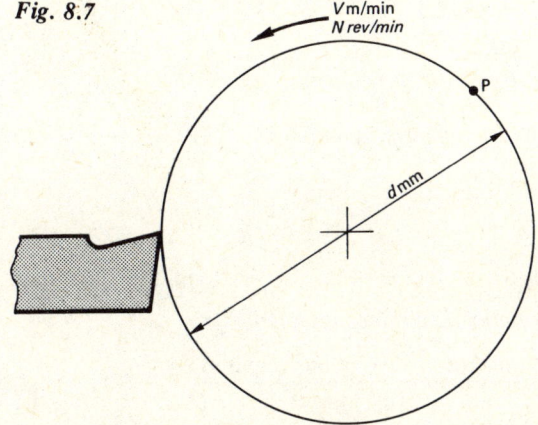

Hence V = cutting speed, m/min
         N = spindle speed, rev/min
         d = workpiece diameter, mm

The cutting speed is the surface speed of the workpiece and may be found by considering the movement of a point P on the workpiece surface.

In one revolution of the workpiece, the point P will move a distance equal to the circumference. Therefore

Distance moved in mm in 1 rev. $= 2\pi \times$ radius

$$= 2\pi \times \frac{\text{diam.}}{2}$$

$$= 2\pi \frac{d}{2}$$

$$= \pi d$$

In one minute the point P will make N revolutions.

$\therefore$ Distance moved in mm in 1 min
     = distance moved in mm in 1 rev $\times N$
     $= \pi d N$

The cutting speed is required in m/min.
Since 1 m = 1 000 mm

Distance moved in m/min $= \dfrac{\pi d N}{1\,000}$

$\therefore \quad V = \dfrac{\pi d N}{1\,000}$

Example 8.35 When machining round bar on a centre lathe, the cutting speed at the periphery of the bar can be found from the formula:

$$V = \frac{\pi d N}{1\,000}$$

where V = cutting speed, m/min
         N = lathe spindle speed, rev/min
         d = diameter of bar, mm.

Use the formula to calculate the cutting speed when machining a bar of 30 mm diameter at a spindle speed of 280 rev/min.

$$N = 280 \text{ rev/min} \qquad d = 30 \text{ mm}$$

$$V = \frac{\pi d N}{1\,000}$$

$$= \frac{22 \times 30 \times 280}{7 \times 1\,000} = 26.4 \text{ m/min} \quad (Ans.)$$

2 Different work materials can be cut most efficiently at different cutting speeds. For instance, aluminium can be machined at a higher cutting speed than carbon steel. The choice of cutting speed also depends on the tool material; a cemented-carbide tool would normally cut at a much higher speed than a high-speed-steel tool.

Generally the cutting tool manufacturer's recommended cutting speed would be used, and the spindle rev/min would be selected to give that speed at the periphery (surface) of the work. Thus, in practice the formula would be used to find N for a known value of V.

The formula must be transposed to make N the subject.

$$V = \frac{\pi d N}{1\,000}$$

$$1\,000V = \pi d N$$

$$\frac{1\,000V}{\pi d} = N$$

$$N = \frac{1\,000V}{\pi d}$$

Example 8.36 Calculate the lathe spindle speed required to machine 20 mm diameter bar at a cutting speed of 33 m/min.

$$V = 33 \text{ m/mm} \qquad d = 20 \text{ mm}$$

$$N = \frac{1\,000}{\pi d}$$

$$= \frac{1\,000 \times 33}{\frac{22}{7} \times 20}$$

$$= \frac{1\,000 \times 33 \times 7}{22 \times 20} = 525$$

Spindle speed $= 525\,\text{rev/min}$ (*Ans.*)

(The nearest available headstock speed to 525 rev/min would be used.)

3   In the drilling and milling operations it is the cutting edge of the tool which is moving past the stationary workpiece. Thus the cutting speed in these operations is the surface speed at the periphery of the tool. The same formula will apply but d represents the diameter of the tool.

Example 8.37   Estimate the correct spindle speed of a drilling machine when using a 10 mm diameter drill at a cutting speed of 35 m/min.

$$V = 35\,\text{m/min} \qquad d = 10\,\text{mm}$$

$$N = \frac{1\,000V}{\pi d} = \frac{1\,000 \times 35}{3.142 \times 10} = 1\,114\,\text{rev/min}   (\textit{Ans.})$$

(The nearest available spindle speed would be used.)

Example 8.38   A milling cutter 100 mm diameter is operating at a spindle speed of 120 rev/min. Calculate the cutting speed.

$$N = 120\,\text{rev/min} \qquad d = 100\,\text{mm}$$

$$V = \frac{\pi d N}{1\,000} = \frac{3.142 \times 100 \times 120}{1\,000} = 37.7\,\text{m/min}   (\textit{Ans.})$$

8.8 Formula for Cutting Time

1   In the lathe operation shown in fig. 8.8, the diameter of the workpiece is being reduced for a distance of L mm along its length. The workpiece is rotating at a spindle speed of N rev/min while the tool moves a distance of f mm along the axis of the work for each revolution. A single cut is taken.

Thus $N = $ spindle speed, rev/min
  $f = $ tool feed, mm/rev
  $L = $ length to be cut, mm.

Let $T = $ time taken, minutes.

The distance moved by the tool in 1 revolution is f, and the total distance to be moved by the tool is L. Therefore

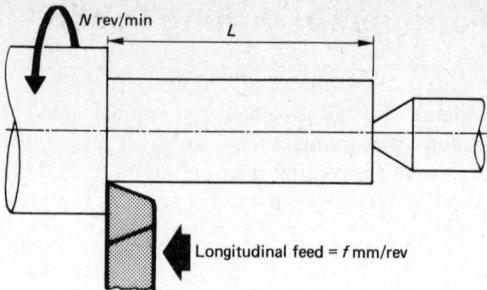

Fig. 8.8

$$\text{Number of revs. required} = \frac{\text{length to be cut}}{\text{feed per revolution}}$$

$$= \frac{L}{f}$$

In 1 minute the work makes N revolutions. Therefore

Number of minutes required

$$= \frac{\text{number of revolutions required}}{\text{rev/min}}$$

$$= \frac{\left(\dfrac{L}{f}\right)}{N}$$

$$= \frac{L}{fN}$$

$$\therefore \quad T = \frac{L}{fN}$$

This formula gives the time taken for the tool to move once along the length L, i.e. the work diameter is reduced in a single cut. If the amount of material to be removed requires a number of cuts, then the total time taken in machining would be given by

$$T \times \text{number of cuts}$$

Example 8.39   Estimate the time taken in minutes in a lathe operation to take one cut of length 150 mm using a longitudinal feed of 0.3 mm/rev and a spindle speed of 200 rev/min.

$$N = 200\,\text{rev/min} \qquad f = 0.3\,\text{mm/rev} \qquad L = 150\,\text{mm}$$

$$T = \frac{L}{fN} = \frac{150}{0.3 \times 200} = 2.5\,\text{min}   (\textit{Ans.})$$

2   Fig. 8.9 shows a facing operation on a lathe. The cutting tool is fed transversely across the face of the workpiece from the circumference to the centre. Thus the length to be cut is $L = $ radius of the work.

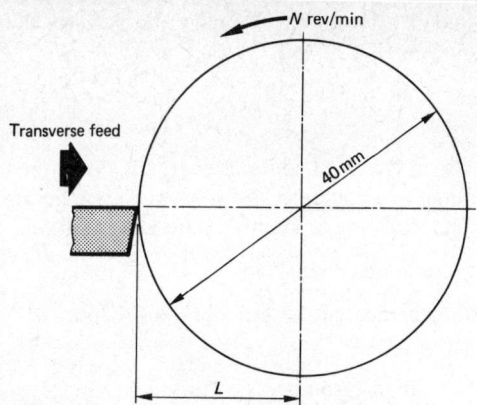

Fig. 8.9

Example 8.40 Find the time taken in minutes to face a 40 mm diameter bar in a single cut on a lathe using a transverse feed of 0.2 mm/rev and a spindle speed of 200 rev/min.

$N = 200 \text{ rev/min} \qquad f = 0.2 \text{ mm/rev}$
$L = \text{radius of bar} = 20 \text{ mm}$

$$T = \frac{L}{fN} = \frac{20}{0.2 \times 200} = 0.5 \text{ min} \quad (Ans.)$$

3 The same formula for cutting time can be applied to the drilling operation (fig. 8.10).

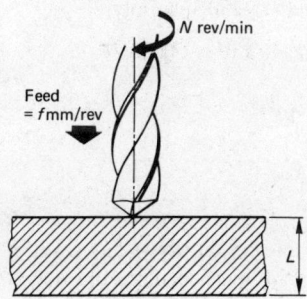

Fig. 8.10

It can be seen from the diagram that, because of the drill point, the actual distance that the drill would feed to drill through the workpiece is slightly greater than L. However, as machining times are normally not required to a high degree of accuracy, the effect of the drill point is usually neglected.

Example 8.41 Estimate the time required in minutes to drill through a plate 60 mm thick using a drill feed of 0.15 mm/rev and a spindle speed of 250 rev/min. (Neglect the effect of the drill point.)

$N = 250 \text{ rev/min} \qquad f = 0.15 \text{ mm/rev} \qquad L = 60 \text{ mm}$

$$T = \frac{L}{fN} = \frac{60}{0.15 \times 250} = 1.6 \text{ min} \quad (Ans.)$$

8.9 Formula for Metal Removal Rate

One method of testing the efficiency of a cutting process is by evaluating its **metal removal rate**, i.e. the volume of metal removed in one minute. A process with a high metal removal rate is generally more economic to use than one with a low rate.

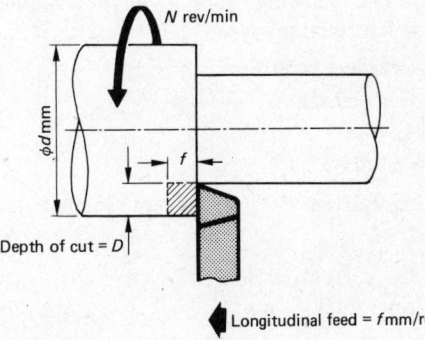

Fig. 8.11

Fig. 8.11 shows a sliding operation on a centre lathe. The area of the chip removed is shown as the shaded rectangle and is given by

Area of chip $= fD \text{ mm}^2$

In 1 revolution, the length of chip removed will be approximately equal to the circumference of the work (neglecting chip deformation).

Length of chip/rev $= 2\pi \times$ radius of the work
$$= 2\pi \times \frac{d}{2}$$
$$= \pi d \text{ mm}$$

The volume of metal removed in 1 revolution
= area of chip $\times$ length of chip/rev
$= fD \times \pi d$
$= \pi d f D \text{ mm}^3$

The volume of metal removed per minute
= volume removed/rev $\times$ rev/min
$= \pi d f D \times N$
$= \pi d N f D \text{ mm}^3/\text{min}$

$\therefore$ Metal removal rate $M = \pi d N f D \text{ mm}^3/\text{min}$

where d = workpiece diameter, mm
N = spindle speed, rev/min
f = longitudinal feed mm/rev
D = depth of cut.

The formula for metal removal rate can also be given in terms of cutting speed.

$$V = \frac{\pi dN}{1\,000}$$

$\therefore \quad 1\,000V = \pi dN$

Substituting in $M = \pi dNfD$

$M = 1\,000VfD$

where V = cutting speed, m/min

Example 8.42 The following data is given for a sliding operation on a centre lathe:

Diameter of workpiece = 50 mm
Longitudinal feed = 0.4 mm/rev
Depth of cut = 8 mm
Spindle speed = 190 rev/min

Calculate the metal removal rate in mm³/min of the process.

$d = 50\,\text{mm} \qquad N = 190\,\text{rev/min}$
$f = 0.4\,\text{mm/rev} \qquad D = 8\,\text{mm}$

$M = \pi dNfD$
$\quad = 3.142 \times 50 \times 190 \times 0.4 \times 8$
$\quad = 99\,520\,\text{mm}^3/\text{min} \quad (Ans.)$

Example 8.43 A turning operation is performed at a cutting speed of 28 m/min with a longitudinal feed of 0.3 mm/rev and a depth of cut of 5 mm. Calculate the metal removal rate in mm³/min.

$V = 28\,\text{m/min} \qquad f = 0.3\,\text{mm/rev} \qquad D = 5\,\text{mm}$

$M = 1\,000VfD$
$\quad = 1\,000 \times 28 \times 0.3 \times 5$
$\quad = 42\,000\,\text{mm}^3/\text{min} \quad (Ans.)$

8.10 Formulae for Electrical Power

1 Electrical energy is discharged when an electric current flows against a resistance. All energy is measured by a common S.I. unit, the joule (symbol J). Power is the rate at which energy is used and is measured by a common S.I. unit, the watt (symbol W).

When 1 joule of energy is consumed in 1 second, the rate of doing work is 1 watt.

1 watt = 1 joule per second
$1\,\text{W} = 1\,\text{J/s}$

In an electrical circuit, 1 joule of energy is discharged when a charge of 1 coulomb passes between two points on a conductor having a potential difference of 1 volt.

Joules = coulombs × volts

An electric current of 1 coulomb per second is 1 ampere.

$\therefore \qquad$ Joules per second = amperes × volts
Watts = amperes × volts

This equation for electrical power can be written:

$P = EI$

where P = power in watts (W)
E = potential difference in volts (V)
I = current in amperes (A)

Example 8.44 An electric lamp draws a current of 0.5 A from a power supply of 240 V. Calculate the power consumed by the lamp.

$E = 240\,\text{V} \qquad I = 0.5\,\text{A}$

$P = EI = 240 \times 0.5 = 120\,\text{W} \quad (Ans.)$

Example 8.45 Calculate the current required by a 2 kW electric fire connected to 250 V supply.

$P = 2\,\text{kW} = 2\,000\,\text{W} \quad (\text{since } 1\,\text{kW} = 1\,000\,\text{W})$
$E = 250\,\text{V}$

$P = EI \quad$ and transposing
$I = \dfrac{P}{E} = \dfrac{2\,000}{250} = 8\,\text{A} \quad (Ans.)$

Example 8.46 An electronic circuit draws a current of 15 mA from a 9 V battery supply. Calculate the power consumed in the circuit.

$E = 9\,\text{V}$
$I = 15\,\text{mA} = \dfrac{15}{1\,000} = 0.015\,\text{A} \qquad (1\,\text{A} = 1\,000\,\text{mA})$

$P = EI$
$\quad = 9 \times 0.015$
$\quad = 0.135\,\text{W} \quad \text{or} \quad 135\,\text{mW} \quad (1\,\text{W} = 1\,000\,\text{mW})$
$\qquad\qquad\qquad\qquad\qquad\qquad\qquad (Ans.)$

2 By using Ohm's Law, the power formula can be expressed in terms of the resistance and the current flowing in a circuit.

From Ohm's Law $E = IR$

$\therefore P = EI = (IR)I$

i.e. $P = I^2R$

where R = resistance in ohms (Ω).

Example 8.47 Calculate the power consumed by a heating coil having a resistance of $25\,\Omega$ when the current flowing is $10\,A$.

$I = 10\,A \qquad R = 25\,\Omega$

$P = I^2R$
$\quad = 10^2 \times 25 = 100 \times 25 = 2\,500\,W$ or $2.5\,kW$ (*Ans.*)

Example 8.48 The power consumed by a resistance of $2.5\,\Omega$ is $0.5\,kW$. Calculate the current flowing.

$P = 0.5\,kW = 500\,W \qquad R = 2.5\,\Omega$

$P = I^2R$ and transposing
$I^2 = \dfrac{P}{R} = \dfrac{500}{2.5} = 200$

$\therefore \quad I = \sqrt{200} = 14.14\,A$ (*Ans.*)

3 The power formula can also be expressed in terms of the resistance and the potential difference.

From Ohm's Law $\quad I = \dfrac{E}{R}$

$\therefore \quad P = EI = E\left(\dfrac{E}{R}\right)$

i.e. $\quad P = \dfrac{E^2}{R}$

Example 8.49 Calculate the power consumed by a resistor of $10\,\Omega$ when connected to a $20\,V$ supply.

$E = 20\,V \qquad R = 10\,\Omega$

$P = \dfrac{E^2}{R} = \dfrac{20^2}{10} = \dfrac{400}{10} = 40\,W$ (*Ans.*)

Example 8.50 An electrical circuit having a resistance of $1.2\,\Omega$ consumes $3\,kW$. Calculate the value of the supply voltage.

$P = 3\,kW = 3\,000\,W \qquad R = 1.2\,\Omega$

$P = \dfrac{E^2}{R}$

Transposing

$E^2 = PR = 3\,000 \times 1.2 = 3\,600$
$E = \sqrt{3\,600} = 60\,V$ (*Ans.*)

Exercises 8

8.1 Express the following statements as algebraic expressions using the symbols:
a = first number $\quad b$ = second number $\quad c$ = third number

  (i) The first number plus the second number minus the third number.
 (ii) The first number multiplied by the third number.
(iii) The sum of the three numbers.
 (iv) The product of the three numbers.
  (v) The first and the third numbers added together and divided by the second number.
 (vi) Twice the first number plus three times the third number.
(vii) Three times the second number minus half the third number.
(viii) Three-quarters of the first number plus half the sum of the other two numbers.
 (ix) The average of the numbers.
  (x) The sum of the squares of the numbers.

8.2 Give the area of each of the shapes shown in fig. 8.12 as an algebraic expression. Simplify where possible.

8.3 The total cost of producing a component is made up of machining costs at £7 per hour, fitting costs at £6 per hour, and inspection costs at £10 per hour.
a) Construct an algebraic expression for the total cost of producing one component.
b) Use the expression to find the total cost of producing one component when:
machining time per component $= 1\frac{1}{2}$ hours
fitting time per component $= 4$ hours
inspection time per component $= \frac{1}{2}$ hour.

8.4 Simplify the following expressions where possible:
a) $3x + 5x + 2x$
b) $8y - 3y + 4y$
c) $5a + 2b - 3a$
d) $6x + 5y - 2x + y$
e) $3a + 4b - 2c$
f) $8x + 4x^2 - 3x + 2x^2 + 5x$
g) $6x + 2x^2 - 4x$
h) $7a + 3a^2 - 4a + 2a^3 - a^2 + 5a$

8.5 Simplify:
a) $3b \times 2b$ *b)* $7a^2 \times 3a$
c) $4y \times 5y^2$ *d)* $2x^2 \times 3x^2$
e) $2a \times 4a \times 5a$ *f)* $4 \times a \times a^2$
g) $3a \times 2b$ *h)* $8x \times 4y \times 3z$

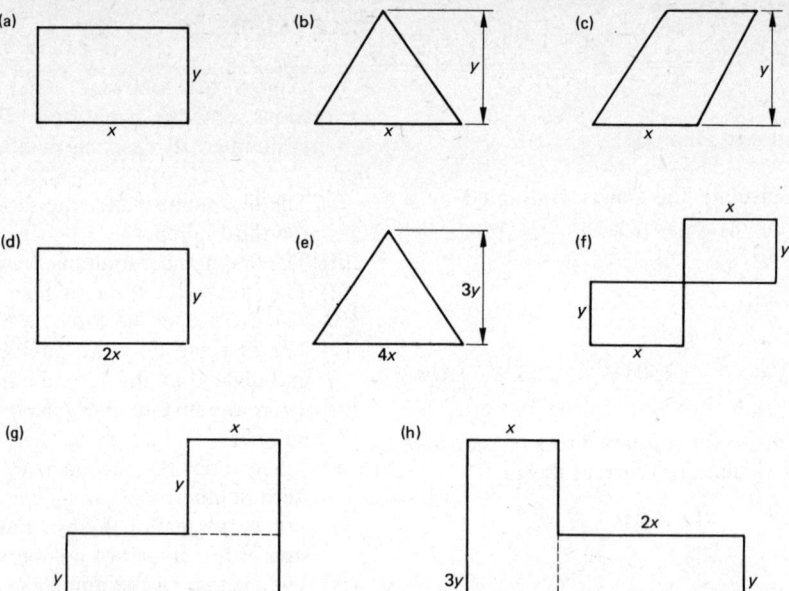

Fig. 8.12

8.6 Simplify:

a) $\dfrac{30ab}{2b}$ b) $\dfrac{24xy}{6x}$ c) $\dfrac{18a^2}{2a}$

d) $\dfrac{108m^3}{9m}$ e) $\dfrac{3a^2b}{ab}$ f) $\dfrac{20a^2c^2}{5ac}$

g) $\dfrac{15x^2y^3}{3xy}$ h) $\dfrac{45a^4x^2}{9a^2x}$

8.7 Expand:

a) $3(a+b)$ b) $2(3x+4)$
c) $4(6a+b)$ d) $10(2x-3y)$
e) $7(4x^2+y)$ f) $3(2a^2-b^2)$
g) $x(1+y)$ h) $4(5a^2-2b^2+c)$
i) $x(x+y)$ j) $y(2x-y)$

8.8 Simplify:

a) $\dfrac{12(x^2+x)}{3x}$

b) $\dfrac{2(6a^2+4a)}{4a}$

c) $\dfrac{9(2a^2-4a)}{6a}$

8.9 Given that $a=2$, $b=3$ and $c=4$, calculate the value of

i) $a+b+c$ ii) $a+b-c$ iii) $2a+2b$
iv) $3b-c$ v) $6a-2b+2c$ vi) $a-c$
vii) $c+(b-a)$ viii) $b-(c+a)$

8.10 Given that $a=4$, $b=2$ and $c=5$, evaluate
i) ab ii) $ac+b$ iii) abc iv) $2a+bc$
v) $a(b+c)$ vi) $c(b-a)$ vii) $\dfrac{2a+b}{c}$ viii) $\dfrac{2bc}{a}$

8.11 Given that $x=2$, $y=-2$ and $z=4$, evaluate
i) $x+y+z$ ii) $2x-y$
iii) $y+2z$ iv) xyz v) $3yz$
vi) $x(y+z)$ vii) $\dfrac{xy}{z}$ viii) $\dfrac{8xz}{y}$

8.12 Given that $m=3$, $n=2$ and $p=4$, evaluate
i) m^2+n^2 ii) $3p^2$ iii) m^3 iv) p^2-n^2
v) $6m^2-n^3$ vi) $\dfrac{mp^2}{n}$ vii) $(m+n)^2$ viii) p^4+m^3

8.13 Given that $K=100$, $L=0.4$ and $M=1.2$, evaluate
i) $KL+M$ ii) $LM+K$
iii) $\dfrac{M}{K}+L$ iv) $\dfrac{MK}{1\,000}$
v) $0.3K+10L$ vi) $\dfrac{3K}{4}-5M$
vii) $\dfrac{M^2+L}{K}$ viii) $\dfrac{30L+5M}{K}$

Solve the Simple Equations, Exercises 8.14 to 8.21

8.14 a) $x + 7 = 10$ b) $x + 16 = 30$
c) $x + 4 = 19$ d) $m + 15 = 20$
e) $y + 8 = 35$ f) $x + 6 = 4$
g) $y + 9 = 6$ h) $x + 1.3 = 2.5$
i) $p + 0.7 = 4.2$ j) $x + 1.318 = 5.75$
k) $x + \frac{1}{4} = 1\frac{1}{2}$ l) $n + \frac{3}{8} = 2\frac{1}{4}$

8.15 a) $x - 6 = 4$ b) $x - 9 = 2$
c) $x - 1 = 5$ d) $x - 13 = 27$
e) $y - 25 = 100$ f) $p - 0.4 = 0.6$
g) $x - 1\frac{3}{4} = 6\frac{1}{2}$ h) $n - 0.318 = 4.264$
i) $y - 7 = -4$ j) $x - 3 = -7$

8.16 a) $10x = 30$ b) $8x = 48$
c) $4x = 32$ d) $7x = 21$
e) $\frac{1}{2}x = 16$ f) $0.3y = 0.9$
g) $1.2x = 6$ h) $1.8x = 12.6$

8.17 a) $\dfrac{x}{5} = 2$ b) $\dfrac{x}{7} = 3$ c) $\dfrac{x}{2} = 1$

d) $\dfrac{x}{10} = 0.2$ e) $\dfrac{x}{20} = 0.3$ f) $\dfrac{p}{9} = 1.61$

g) $\dfrac{m}{3.21} = 4$ h) $\dfrac{x}{0.01} = 0.1$

8.18 a) $2x + 1 = 7$ b) $4x + 2 = 18$
c) $5p - 1 = 9$ d) $3x - 8 = -11$
e) $2p - 9 = -8$ f) $5x - 16 = 34$
g) $7m - 12 = 72$ h) $2x - \frac{1}{2} = 2$

8.19 a) $\frac{1}{2}x + 6 = 8$ b) $\frac{3}{4}x - 1 = 5$
c) $\frac{2}{3}m + 6 = 12$ d) $\frac{4}{3}K + 0.6 = 1$

8.20 a) $3(x + 2) = 18$ b) $7(x - 6) = 14$
c) $2(x + \frac{1}{2}) = 6$ d) $3(4x - 1) = 57$
e) $8(2x + 4) = 64$ f) $2(x + 1) + 3(x - 1) = 19$
g) $3(m - 1) + 2(m + 2) = 36$
h) $5(p + 2) + 2(p + 5) = 62$

8.21 a) $3(x + 9) = 2(x + 5) + 27$
b) $8(p + 6) = 7(p - 1) + 54$
c) $3(k - 2) = 4(k + 3) - 24$
d) $2(x + 0.6) = 3(x + 1) - 3.8$
e) $10(x + 1.1) = 4(x - 0.7) + 30$
f) $5(x + 1.2) = 2(x + 0.7) + 5.5$

8.22 Transpose:
 i) $F = xyt$ for t
 ii) $g = mh$ for m
 iii) $s = \pi r h$ for r
 iv) $I = PRT$ for T
 v) $A = \pi r^2 v$ for v
 vi) $V = \dfrac{\pi D^2 l}{4}$ for l
 vii) $n = pm - q$ for m
 viii) $\dfrac{M}{I} = \dfrac{E}{R}$ for R

8.23 Transpose:
 i) $y = a(x + b)$ for x
 ii) $C = 100(k - a)$ for k
 iii) $M = p(n - 2)$ for n
 iv) $t = r + s(n - 1)$ for n
 v) $G = \dfrac{x}{p + q}$ for x
 vi) $x = \dfrac{0.9}{b + d}$ for b

8.24 The total resistance of two electrical resistors wired in parallel can be found from the formula:

$$\frac{1}{R} = \frac{1}{R_1} + \frac{1}{R_2}$$

where R = total resistance, ohms
R_1 = first resistor, ohms
R_2 = second resistor, ohms.

Use the formula to find the total resistance of two resistors of 12 ohms and 6 ohms connected in parallel.

8.25 Temperature in degrees Celsius may be converted to degrees Fahrenheit by use of the formula:

$$F = \frac{9C}{5} + 32$$

where C = degrees Celsius
F = degrees Fahrenheit.

a) Convert to degrees Fahrenheit the melting point of a brazing spelter given as 900° Celsius.
b) Produce a formula by transposition to convert degrees Fahrenheit to degrees Celsius.

8.26 The relationship between voltage V, resistance R and current flowing I in an electrical circuit is given by
$$V = IR$$
with V in volts, I in amps, R in ohms.
Calculate the value of I when $V = 250$ volts and $R = 40$ ohms.

8.27 a) Using the symbols A, b and h, construct a general formula for the area of a triangle.
b) Using the symbols V, r and l, construct a general formula for the volume of a solid cylinder.

8.28 Fig. 8.13 shows the angles on the face of a lathe tool.

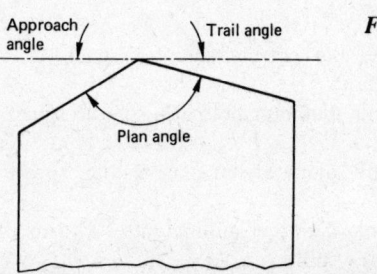

Approach angle   Trail angle   *Fig. 8.13*

Plan angle

a) Construct a formula to give the approach angle in terms of the other two angles.
b) Determine the plan angle when the approach angle = 35° and the trail angle = 12°.

8.29 A piece of pipe is bent to the shape shown in fig. 8.14.

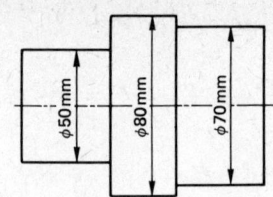

Fig. 8.15

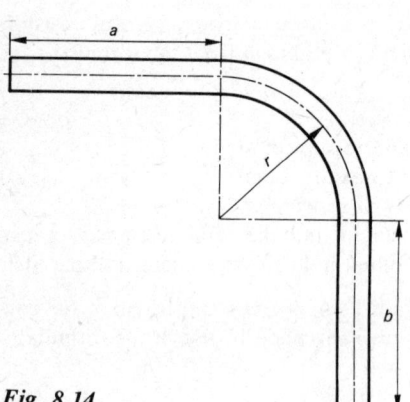

Fig. 8.14

a) Construct a formula to find *L* the overall length of the pipe.
b) Find the length of the pipe when *a* = 80 mm, *b* = 65 mm, *r* = 35 mm.

8.30 Calculate the spindle speed required when
a) turning 30 mm diameter bar at 26 m/min
b) drilling a 10 mm diameter hole at 40 m/min
c) using a 100 mm diameter milling cutter at 32 m/min
d) boring a 40 mm diameter hole at 20 m/min.

8.31 *a*) Construct a formula to find the speed in m/min at the periphery of a grinding wheel of diameter *d* mm.
b) Use the formulae to find the surface speed of a grinding wheel having a diameter of 150 mm and a spindle speed of 3 000 rev/min.

8.32 Determine the cutting speed in each of the following operations:
a) drilling a 5 mm diameter hole at a spindle speed of 1 800 rev/min
b) turning 15 mm diameter bar at a spindle speed of 580 rev/min
c) using a 150 mm diameter milling cutter at a spindle speed of 52 rev/min.

8.33 The shaft shown in fig. 8.15 is rotated in a lathe at a spindle speed of 150 rev/min. Calculate the surface speed in m/min of each of the three diameters.

8.34 Estimate the time taken in minutes in a lathe operation to take one cut of length 250 mm using a longitudinal feed of 0.25 mm/rev and a spindle speed of 300 rev/min.

8.35 *a*) Construct a formula using the following symbols to give the time taken to drill a hole. (Neglect the effect of drill point.)

Feed	$= f$ mm/rev
Spindle speed	$= N$ rev/min
Depth of hole	$= L$ mm
Time taken	$= T$ min

b) Find the time taken in seconds to drill a hole 30 mm deep using a spindle speed of 400 rev/min and a feed of 0.2 mm/rev.

8.36 The following information is given for a drilling operation:

Drill diameter	$= 6$ mm
Depth of hole	$= 40$ mm
Cutting speed	$= 30$ m/min
Drill feed	$= 0.25$ mm/rev

a) Calculate the spindle speed in rev/min.
b) Calculate the time taken in seconds to drill the hole. (Neglect the effect of drill point.)

8.37 Calculate the time taken in minutes and seconds to face a diameter of 120 mm in a single cut on a lathe. The transverse feed is 0.25 mm/rev and the spindle speed is 100 rev/min.

8.38 Given that $V = E - IR$
a) Transpose the formula to give
i) *E* ii) *I* iii) *R*
b) Find the value of
i) *V* when $E = 3$, $I = 1$ and $R = 0.7$
ii) *I* when $V = 5.1$, $E = 6$ and $R = 0.6$
iii) *R* when $V = 9.6$, $E = 12$ and $I = 3$
iv) *E* when $V = 7.2$, $I = 2.5$ and $R = 0.75$

8.39 Use Ohm's Law $E = IR$ to complete the following table.

E (volts)	I (amps)	R (ohms)
	5	10
240	2.5	
12		6.5
	3	35
24	0.6	
71		20

8.40 Given that $P = I^2R$
a) Transpose the formula to give   i) R ii) I
b) Find the value of
 i) P when $I = 0.5$ and $R = 480$
 ii) I when $P = 500$ and $R = 3.5$
 iii) R when $P = 33.75$ and $I = 1.5$

8.41 Given that $P = EI$, use Ohm's Law to derive the formulae:

a) $P = I^2R$ b) $P = \dfrac{E^2}{R}$

8.42 Calculate the power consumed by
a) a heating element taking a current of 4.5 A from 240 V supply
b) an electric lamp taking a current of 0.4 A from 250 V supply
c) an electronic circuit taking a current of 20 mA from 6 V supply.

8.43 An electric lamp consumes 75 W from 240 V supply. Calculate a) the current flowing, b) the resistance of the lamp.

8.44 An electromotive force of 24 V produces a current of 1.8 A in a circuit. Calculate
a) the resistance of the circuit
b) the power consumed
c) the e.m.f. required to produce a current of 2.5 A in the same circuit.

8.45 Calculate the power consumed in kilowatts by
a) a resistor of 7.5 Ω connected to a 20 V supply
b) an electrical circuit having a resistance of 1.2 Ω connected to a 50 V supply
c) a resistance of 2.5 Ω connected to a 500 V supply.

9 Graphs

Graphs are a means of presenting numerical information in a simple pictorial form so that it may be readily assessed and interpreted. Large amounts of data can be shown in a very concise manner that allows the situation to be viewed as a picture, showing clearly how one quantity is changing with respect to another. Graphs are used to make it easier to understand information, and, once the scale of a graph is known, its picture of the information means the same in any language. Hence, graphs are much used in international technical communications.

9.1 Presenting Information by Graphs

The table below shows the number of finished parts of the same type produced in a machine shop in a production run of 16 consecutive working weeks.

W	1	2	3	4	5	6	7	8
FP	521	744	912	952	938	951	706	549
W	9	10	11	12	13	14	15	16
FP	363	385	482	565	781	945	938	951

The table gives numerical information about what is happening in the practical situation of the machine shop, but it is difficult to understand the information in this form because of the large number of figures. Fig. 9.1 shows the same numerical information in the form of a graph. Each point on the graph represents one week's production, and the points are joined by straight lines to give a picture of the information.

The graph clearly shows that, at the beginning of the run, the production rate increased rapidly to remain steady at a high level, then dropped dramatically to a very low level. Over the remaining weeks of the run, the production rate climbed again and finally remained steady at its previous high level. This is a lengthy and complicated description of what has taken place, but all this can be seen in one glance at the graph.

The picture given by the graph can be interpreted to give important information about the practical situation:

1) The process takes some time after its introduction to achieve a high production rate (weeks 1 to 3).
2) The process is capable of maintaining steady production at a high level (weeks 3 to 6 and weeks 14 to 16).

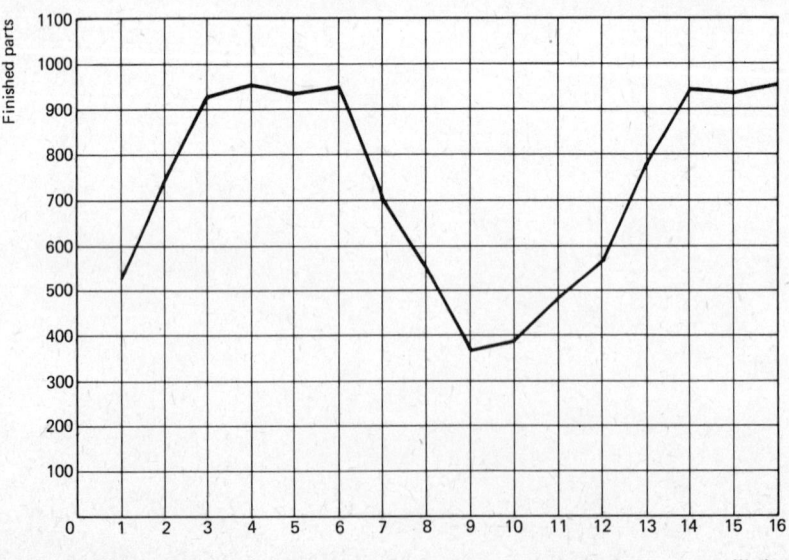

Fig. 9.1

3) Some adverse factor affected production (weeks 6 to 9) and reduced it to a very low level. Investigation may show that factor to be a temporary shortage of materials, tools or labour.
4) The temporary problem was overcome and production was gradually increased back to its previous high level (weeks 9 to 14).

Example 9.1 The table below gives measurement of the temperature of a small furnace taken at hourly intervals from the time it was switched on at 09.00 until 16.00. Fig. 9.2 shows this data in the form of a graph. What information can be interpreted from the graph?

time							
09.00	10.00	11.00	12.00	13.00	14.00	15.00	16.00
20	275	565	575	580	575	495	360
temperature °C							

1) The furnace was heating up at a fairly uniform rate from room temperature for the first two hours.
2) A steady working temperature of about 575°C was held for a period of 3 hours.
3) The furnace was switched off after 14.00 and left to cool.

Example 9.2 Samples were taken at regular intervals from a large batch of shafts, of nominal diameter 15 mm, being produced on an automatic lathe. The average diameter of the shafts in each sample is given in the table below, and the same data is shown on the graph (fig. 9.3). Explain what information about the process can be interpreted from the graph.

sample number									
1	2	3	4	5	6	7	8	9	10
14.97	14.99	15.03	15.05	15.07	14.95	14.97	15.00	15.02	15.04
average diameter mm									

1) The tooling has been initially set to produce shafts slightly below the nominal 15 mm diameter.
2) As the process continued, the shafts produced were increasing in size; this was because the cutting tool was wearing.
3) After the 5th sample the tool was changed and reset to produce shafts slightly below the nominal size.
4) As the process continued, the tool began to wear and the sample diameter again increased in size.

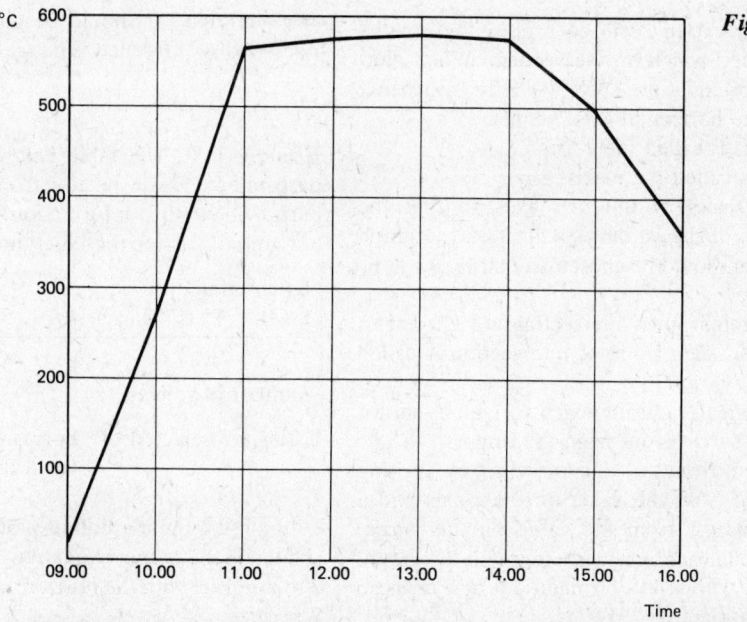

Fig. 9.2

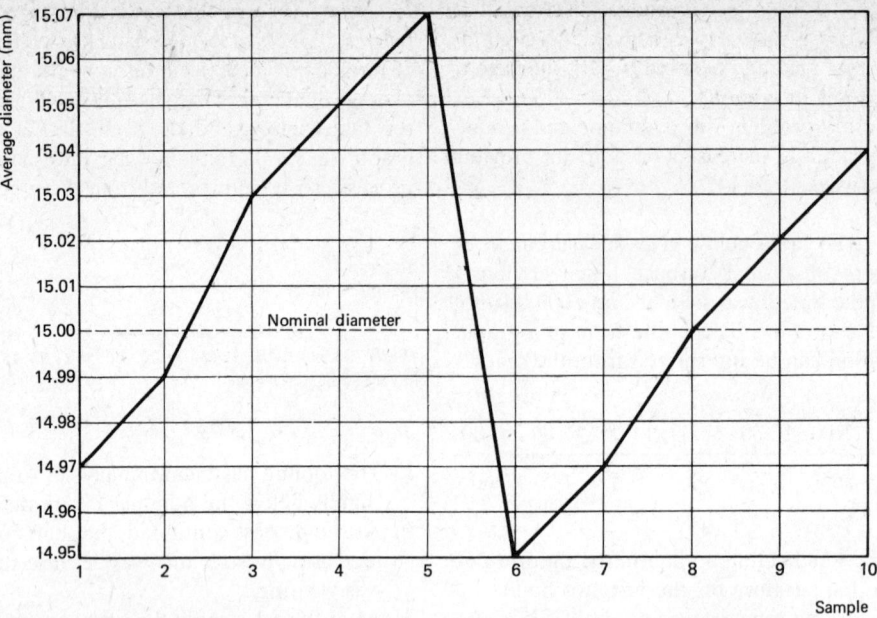

Fig. 9.3

9.2 Plotting Graphs

General procedure for **plotting graphs**:

1 Decide which variable will be represented on the horizontal axis and which on the vertical. When plotting the graphs of equations such as $y = 2x + 3$, then x always lies on the horizontal axis, hence:

Horizontal axis is called the x-axis.
Vertical axis is called the y-axis.

2 Decide the **scales** of the two axes so that the maximum values of the variables will fit on the graph paper. The scales should be chosen so that they will be easy to subdivide.

3 Draw two straight lines intersecting at right-angles to form the axes. Their point of intersection is called the **origin**.

4 Label each axis to indicate which variable is shown and number the subdivisions from the origin.

5 Plot each point from the table of values by drawing a vertical line up from the value on the x-axis and a horizontal line across from the value on the y-axis. Where these two lines intersect is a point on the graph. Mark each point with a dot, a ringed dot or a cross so that it stands out plainly.

6 Join the points with straight lines, or, if the points appear to lie on a single curve, join them with a smooth continuous line to form the curve.

7 If the graph is drawn as part of an investigation or an experiment, it should be given a short descriptive title for easy reference. The scales may also be shown.

Example 9.3 The table below shows the number of parts rejected by inspection from 8 batches of machined parts. Show this information on a graph using the horizontal axis for the batch number.

batch number							
1	2	3	4	5	6	7	8
5	7	12	4	6	11	9	3

number of rejects

Scales are selected of   horizontal   1 batch = 2 cm
                          vertical   1 reject = 1 cm

In fig. 9.4 the method of plotting the second point is shown by the vertical line drawn up from batch number 2 to intersect with the horizontal line drawn across from 7 rejects.

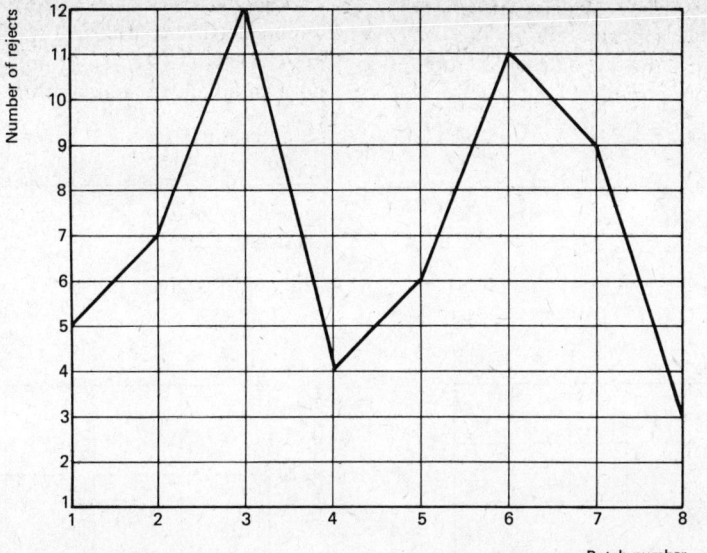

Fig. 9.4

Example 9.4 The table below shows the area of squares having lengths of side from 1 cm to 8 cm. Draw a graph of this information using the horizontal axis for the length of side.

length of side cm							
1	2	3	4	5	6	7	8
1	4	9	16	25	36	49	64
area cm²							

In fig. 9.5 the points are joined by a smooth curve. There could be squares of less than 1 cm side, so the curve may be drawn to zero (the origin).

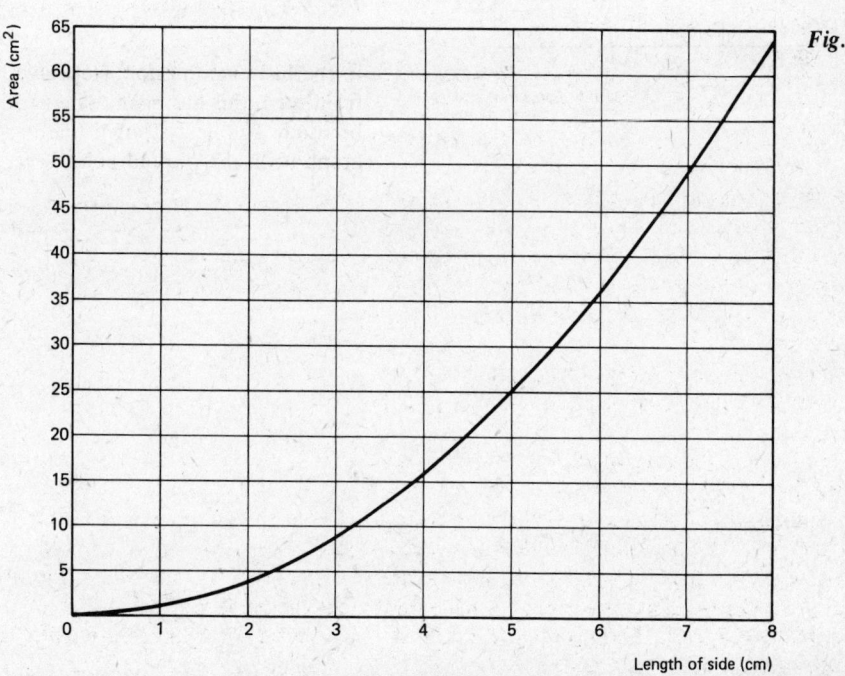

Fig. 9.5

Example 9.5 The table shows the electrical power P in watts taken by an electrical circuit for various values of current I in amperes. Show this data on a graph using the horizontal axis for current.

I amperes							
0	1	2	3	4	5	6	7
0	2	8	18	32	50	72	98
P watts							

The graph is given in fig. 9.6.

Example 9.6 The table shows the resistance in ohms of an electrical conductor over a range of temperatures in °C. Draw a graph of this information using the horizontal axis for temperature.

temperature °C					
20	30	40	50	60	70
6.30	6.70	7.00	7.47	7.80	8.16
resistance Ω					

The graph is given in fig. 9.7.

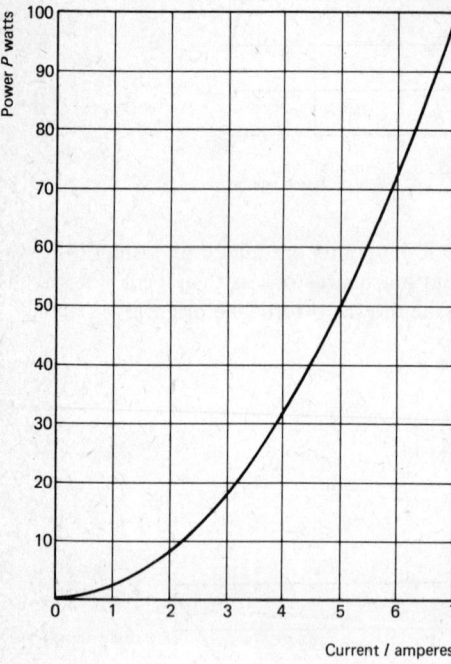

Fig. 9.6

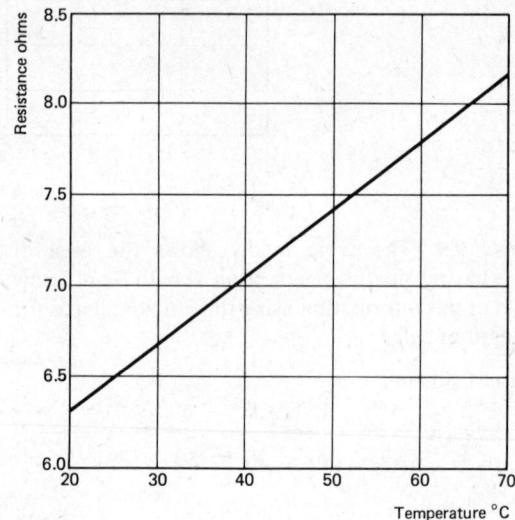

Fig. 9.7

In the last example the axes have not been numbered from zero; this has been done to allow a larger scale to be used. The best straight line has been drawn on the graph to fit the plotted points.

9.3 Interpolation

When the points of a graph produce a definite shape such as the curve in fig. 9.5 or a continuous straight line, then a fixed relationship exists between the two variables.

Consider two variables x and y which are linked by the relationship, $y = 2x$. For a range of values of x, a table can be drawn up to give the values of y:

x	0	1	2	3	4	5
$y = 2x$	0	2	4	6	8	10

This data can be shown on a graph (fig. 9.8)

The graph can be used to find the value of y for *any* value of x within its range.

For example, to find y when $x = 3.5$:

On fig. 9.8, a vertical line is drawn up from the point $x = 3.5$ on the x-axis until it cuts the sloping line of the graph. From this intersection, a horizontal line is drawn across to the y-axis and the value of y is read from the scale.

Thus, when $x = 3.5$, $y = 7$.

This process of finding values within the graph is called **interpolation**.

The value of x for a given value of y may also be found by interpolation.

For example, to find x when $y = 8.5$:

The dotted line on fig. 9.8 is drawn from $y = 8.5$ to cut the line of the graph and then dropped to the x-axis, when the value of x is read from the scale.

Thus, when $y = 8.5$, $x = 4.25$.

Example 9.7 Draw the graph of $y = x^3$, for values of x from 0 to 5. By interpolation from the graph, deduce
a) the value of y when $x = 4.5$
b) the value of x when $y = 60$.

x	0	1	2	3	4	5
$y = x^3$	0	1	8	27	64	125

From fig 9.9.
a) When $x = 4.5$, $y = 91$ (*Ans.*)
b) when $y = 60$, $x = 3.9$ (*Ans.*)

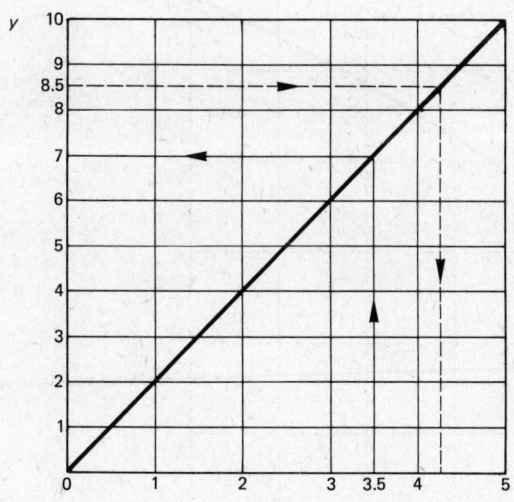

Fig. 9.8

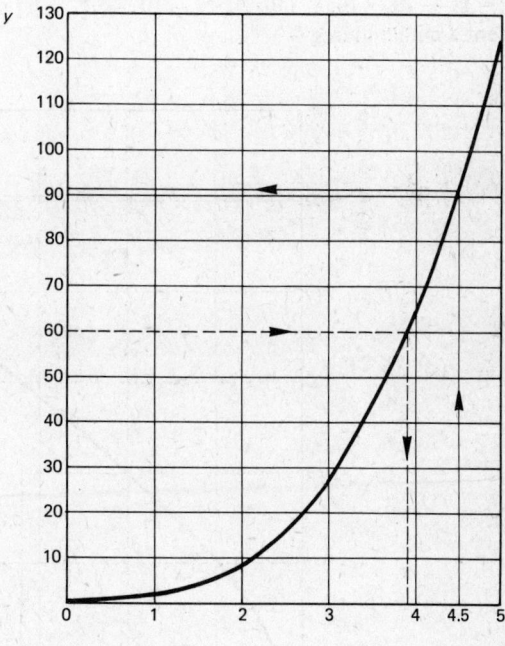

Fig. 9.9

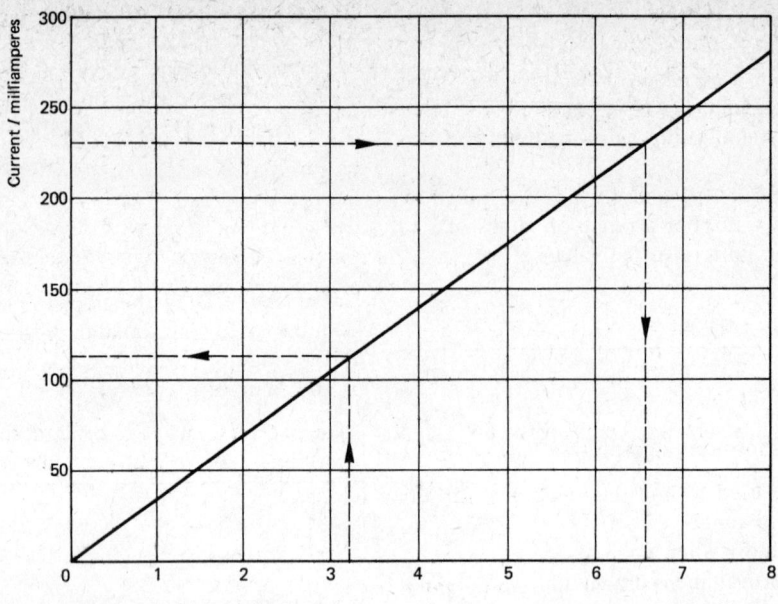

Fig. 9.10

Example 9.8 The following results were obtained for a resistor in an electrical circuit.

voltage V volts								
0	1	2	3	4	5	6	7	8
0	35	70	105	140	175	210	245	280
current I milliamperes								

a) Plot this information on a graph using the horizontal axis for voltage.
b) Interpolate the graph to find:
i) the current corresponding to a voltage of 3.2 volts
ii) the voltage corresponding to a current of 230 milli-amperes.

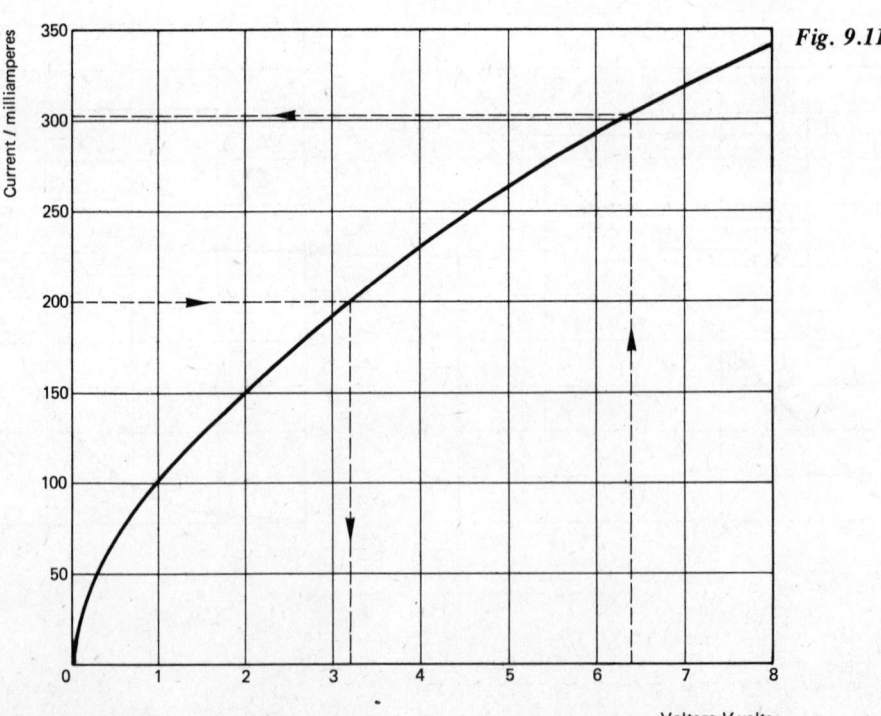

Fig. 9.11

a) See fig. 9.10.
b) From the graph (dotted lines),
 i) When voltage = 3.2 V, current = 112 mA (*Ans.*)

ii) When current = 230 mA, voltage = 6.57 V (*Ans.*)

Example 9.9 The table shows the voltage *V* volts across a lamp when the current flowing is *I* milliamperes.

voltage *V* volts								
0	1	2	3	4	5	6	7	8
0	100	150	190	231	262	290	317	340

current *I* milliamperes

a) Plot this information on a graph using the horizontal axis for voltage.
b) Find i) the current when the voltage is 6.4 volts, ii) the voltage when the current is 200 milliamperes.

a) See fig. 9.11.
b) From the graph (dotted lines),
 i) When voltage = 6.4 V, current = 305 mA (*Ans.*)

ii) When current = 200 mA, voltage = 3.2 V (*Ans.*)

(In this example the relationship between voltage and current is a curve rather than a straight line because the resistance of the lamp is not constant.)

9.4 Conversion Graphs

Graphs are a useful means of rapid conversion from one quantity to another. They are often more convenient than conversion tables because all values can be found within the range of the graph by the use of interpolation. However, the accuracy of the answer will depend on the accuracy of drawing and reading the graph. This section shows some examples of **conversion graphs** and their use by interpolation.

Example 9.10 Draw a graph to convert inches to centimetres for the range 0 to 6 inches. Use the graph to
a) convert 2.5 inches to centimetres
b) convert 12 centimetres to inches.

(1 inch = 2.54 cm)

A table is first constructed by multiplying inches by 2.54.

inches						
0	1	2	3	4	5	6
0	25.4	5.08	7.62	10.16	12.7	15.24

centimetres

From fig.9.12
a) 2.5 inches = 6.35 cm (*Ans.*)
b) 12 cm = 4.75 inches (*Ans.*)

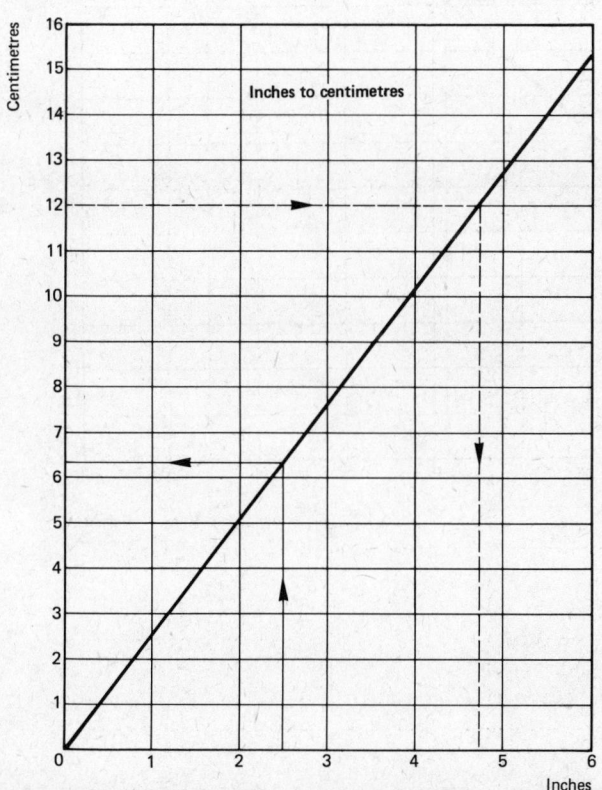

Fig. 9.12

Example 9.11 Draw a graph to convert temperature in degrees Celsius to degrees Fahrenheit for the range 0 to 80°C. Use the graph to
a) convert 38°C to Fahrenheit
b) convert 150°F to Celsius.

$(F = \frac{9}{5}C + 32)$

A table is first constructed using the formula. For example

To convert 0°C to °F

$$°F = \left(\frac{9}{5} \times 0\right) + 32 = 32$$

To convert 10°C to °F

$$°F = \left(\frac{9}{5} \times 10\right) + 32 = 18 + 32 = 50$$

Celsius								
0	10	20	30	40	50	60	70	80
32	50	68	86	104	122	140	158	176
Fahrenheit								

From fig. 9.13
a) 38°C = 100.5°F (*Ans.*)
b) 150°F = 65.5°C (*Ans.*)

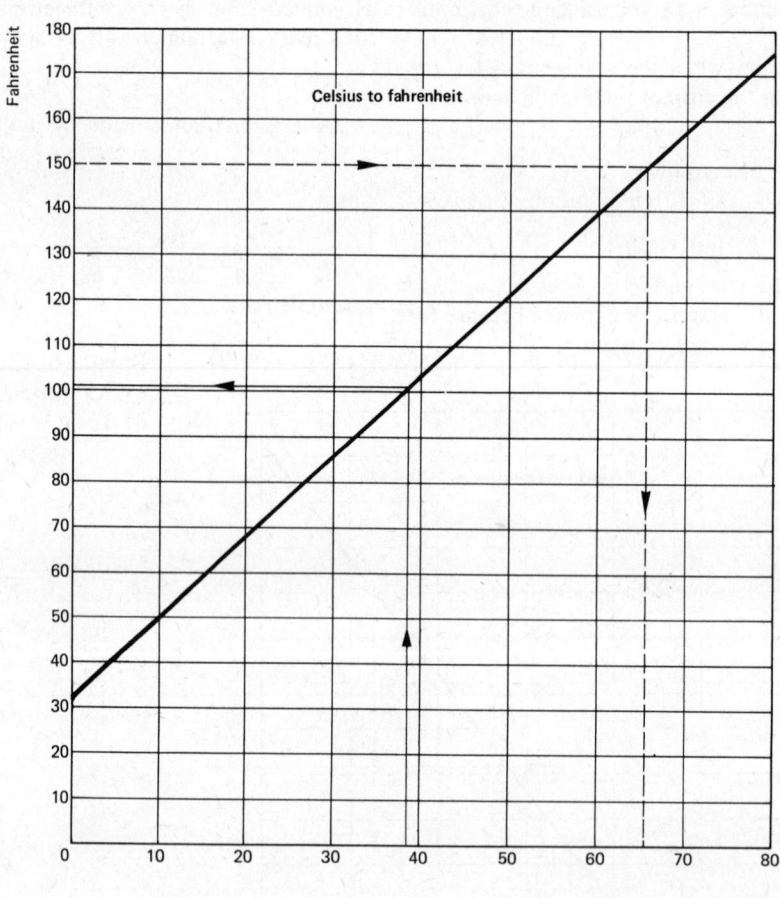

Fig. 9.13

Example 9.12 Draw a graph to show the correct spindle speed in rev/min when machining, on a lathe, a range of work diameters from 20 mm to 80 mm at a cutting speed of 50 m/min. Use the graph to provide
a) the spindle speed for a work diameter of 65 mm
b) the work diameter that would require a spindle speed of 450 rev/min.

$$\left(N = \frac{1\,000\,V}{\pi d}\right)$$

A table is first constructed using the formula. For example, to find the spindle speed for 20 mm diameter:

$$N = \frac{1\,000\,V}{\pi d} = \frac{1\,000 \times 50}{3.142 \times 20} = 796 \text{ rev/min}$$

work diameter d mm						
20	30	40	50	60	70	80
796	530	398	318	265	227	199
spindle speed N rev/min						

From fig. 9.14
a) Spindle speed = 245 rev/min (*Ans.*)
b) Work diameter = 35.5 mm (*Ans.*)

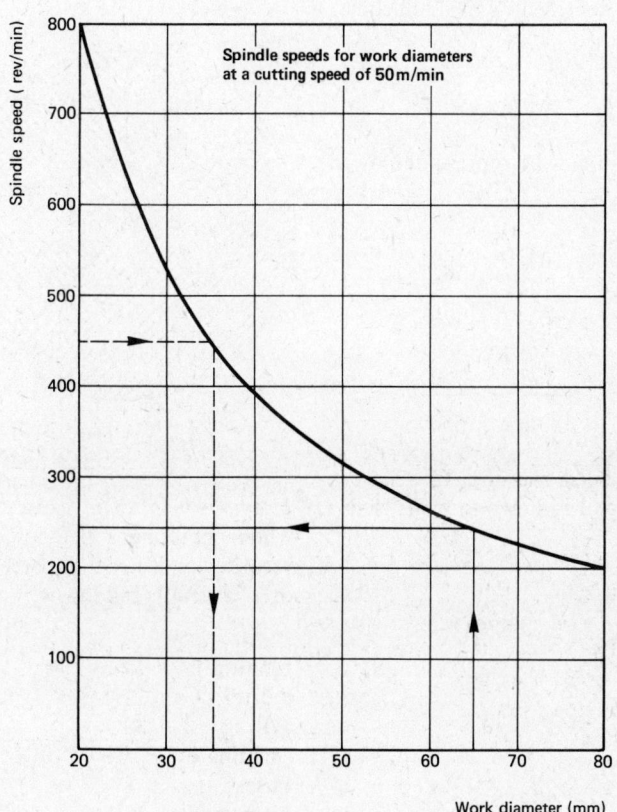

Fig. 9.14

9.5 Presenting Information by Diagrams

1 PIE CHARTS It is often useful to present statistical data or information in the form of a diagram which shows the facts in a clear and simple fashion. One such method of data presentation is the CIRCULAR DIAGRAM or PIE CHART. In this diagram the total of the given information is represented by a full circle, and various sections of the information can be shown as sectors of the circle.

Example 9.13 A small factory has a total labour force of 180 workers. This total is made up of 90 skilled workers, 60 unskilled workers and 30 apprentices. Show this information on a Pie Chart.

The total of 180 workers is represented by a full circle.
∴   180 workers = 360°

$$1 \text{ worker} = \frac{360}{180} = 2°$$

Each section of workers can now be represented by an angle.

90 skilled workers are represented by	$90 \times 2° = 180°$
60 unskilled workers are represented by	$60 \times 2° = 120°$
30 apprentices are represented by	$30 \times 2° = \underline{60°}$
	$180 \times 2° = 360°$

The information can now be shown on a Pie Chart in which the angle of each sector is drawn to scale using a protractor (fig. 9.15).

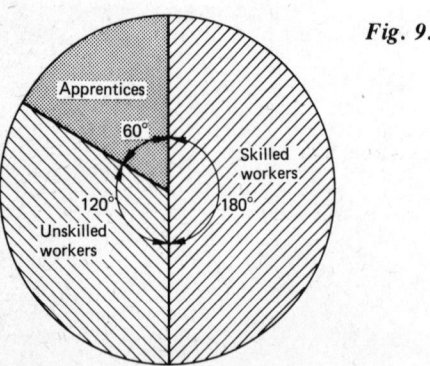

Fig. 9.15

Example 9.14 The total time taken to manufacture a component is made up as follows:

Turning   45 min
Milling   65 min
Drilling   10 min

Show this information on a Pie Chart.

Total time taken = 45 + 65 + 10 = 120 min
120 min = 360°

$$1 \text{ min} = \frac{360}{120} = 3°$$

Turning	$45 \times 3° = 135°$
Milling	$65 \times 3° = 195°$
Drilling	$10 \times 3° = \underline{30°}$
	$120 \times 3° = 360°$

The information can now be shown on a Pie Chart (fig. 9.16).

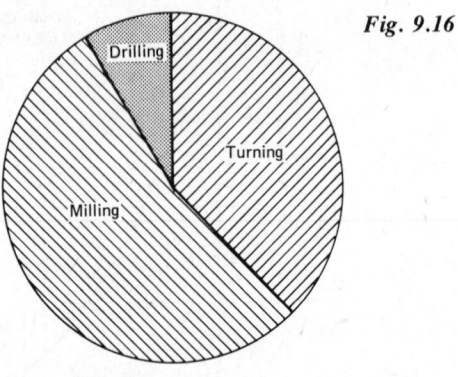

Fig. 9.16

Example 9.15 The causes of electrical breakdowns when operating a fleet of commercial vehicles were recorded and shown as percentages of the total number of electrical breakdowns.

Ignition	10%
Starter	12%
Wiring	30%
Lighting	5%
Fuses	8%
Instruments	15%
Battery	20%

Show this information on a Pie Chart.

The total of 100% is represented by a full circle.

100% = 360°

$$1\% = \frac{360}{100} = 3.6°$$

Ignition	$10 \times 3.6 =$	$36°$
Starter	$12 \times 3.6 =$	$43.2°$
Wiring	$30 \times 3.6 =$	$108°$
Lighting	$5 \times 3.6 =$	$18°$
Fuses	$8 \times 3.6 =$	$28.8°$
Instruments	$15 \times 3.6 =$	$54°$
Battery	$20 \times 3.6 =$	$72°$
	$100 \times 3.6 =$	$360°$

The information can now be shown on a Pie Chart (fig. 9.17).

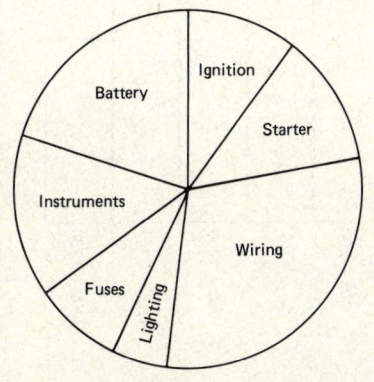

Fig. 9.17

2 BAR CHARTS AND HISTOGRAMS Pie Charts have the disadvantage that a certain amount of calculation and accurate measurement of angle is involved. BAR CHARTS are an alternative method of showing the same information on graph paper.

Example 9.16 A certain component can be produced on 4 alternative types of lathe. The production rate in components per hour for each machine is shown below.

Capstan lathe	18
Centre lathe	8
Turret lathe	15
Automatic lathe	24

Show this information on a Bar Chart.

Fig. 9.18 shows the Bar Chart. The vertical axis is marked with a scale from 0 to 25 to represent the number of components produced per hour. The height of each bar represents the production rate of a particular machine. The horizontal axis has no scale; the bars are equally spaced at some distance apart for the sake of clarity.

A HISTOGRAM has scales on both the horizontal and the vertical axes and is used to present data which applies to a continuous variable.

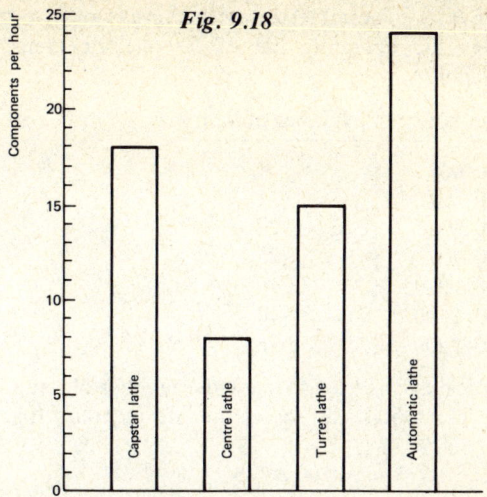

Fig. 9.18

Example 9.17 The table shows the number of twist drills of various sizes held in stock by a machine shop store.

Drill diameter, mm							
3	4	5	6	7	8	9	10
12	16	20	14	8	24	3	30
Number of drills							

Show this information on a Histogram.

The vertical axis is used to show the number of drills and the horizontal axis shows the drill size (fig. 9.19).

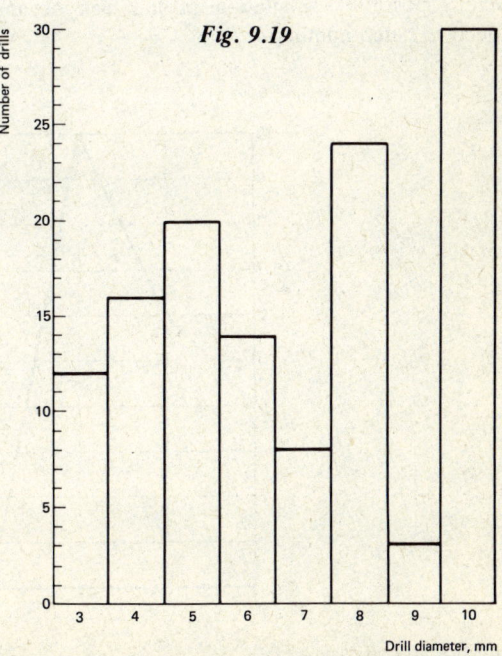

Fig. 9.19

Drill diameter, mm

Graphs 123

Example 9.18 The table shows the results of a test measuring the operating life of 100 electric lamps before failure.

Life in hours	Number of lamps
500–600	4
600–700	14
700–800	20
800–900	25
900–1 000	19
1 000–1 100	10
1 100–1 200	8

The vertical axis is used to show the number of lamps and the horizontal axis shows the life in hours (fig. 9.20).

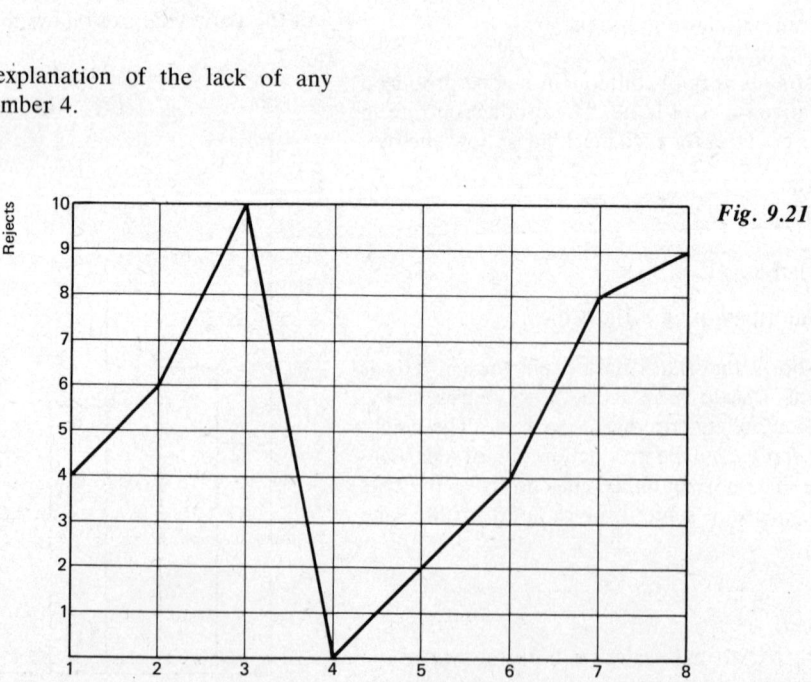

Fig. 9.20

Exercises 9

9.1 Fig. 9.21 shows a graph of the number of rejected parts in 8 batches of machined parts.

a) Use the graph to complete the following table.

batch number

1	2	3	4	5	6	7	8
4							

number of rejects

b) Give a possible explanation of the lack of any rejects in batch number 4.

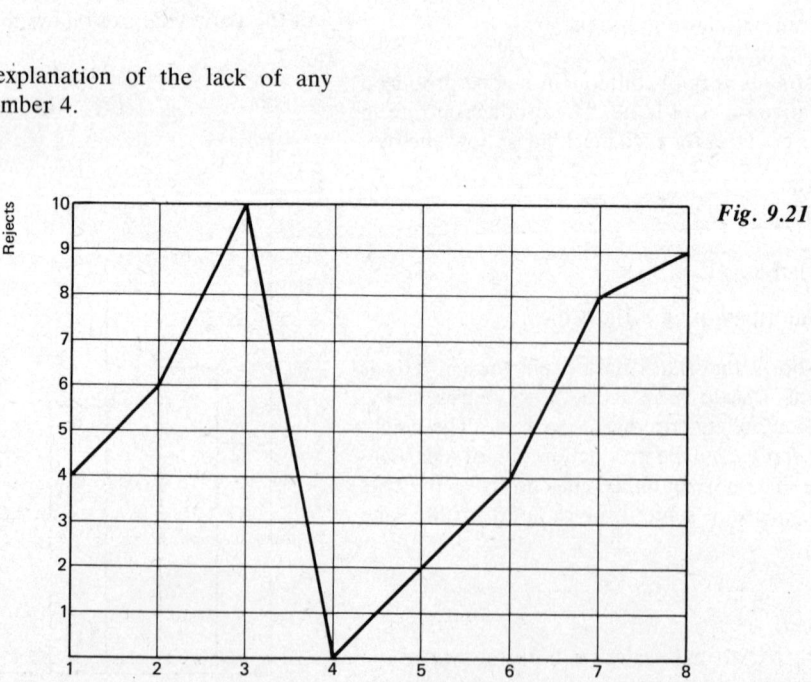

Fig. 9.21

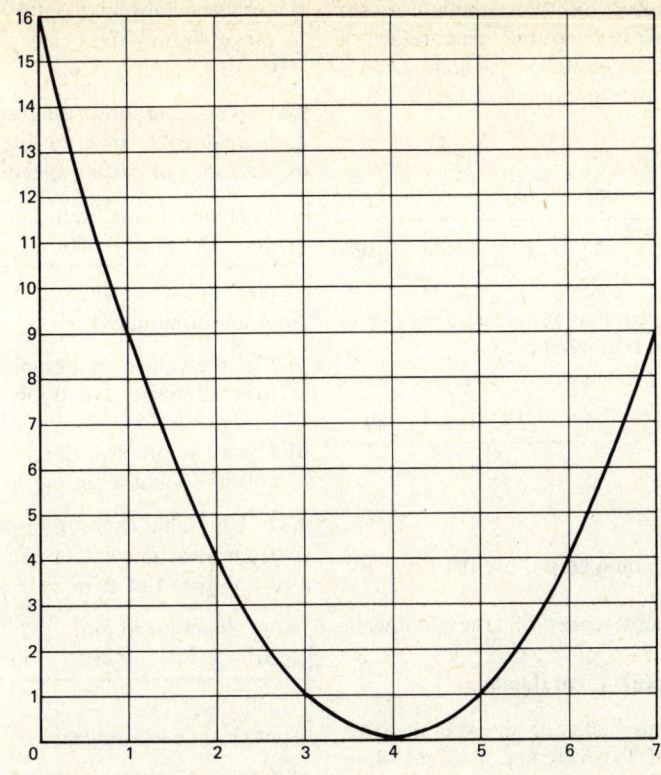

Fig. 9.22

9.2 Use the graph shown in fig. 9.22 to

a) Complete the following table:

x	0	1	2	3	4	5	6	7
y								

b) Estimate from the graph:
 (i) the value of y when $x = 1.5$
 (ii) the value of x when $y = 10$.

9.3 Show the information given in each of the following tables in the form of a graph. The first quantity in each case is to be on the horizontal axis. Use a common scale of $1\,\text{cm} = 1$ on both axes.

a)

x	0	1	2	3	4	5	6	7	8	9
y	0	1	2	3	4	5	6	7	8	9

b)

s	0	1	2	3	4	5	6	7	8
t	2	3	4	5	6	7	8	9	10

c)

m	0	1	2	3	4	5	6	7	8	9
p	9	8	7	6	5	4	3	2	1	0

d)

A	0	1	2	3	4	5	6
B	9	4	1	0	1	4	9

9.4 The table shows the output of a factory in finished units over a period of 9 months. Show this information on a graph using the horizontal axis for months. Give any interpretation that can be made on the shape of the graph.

month								
May	June	July	Aug.	Sept.	Oct.	Nov.	Dec.	Jan.
120	132	135	45	90	130	135	65	95
output								

9.5 The table shows the number of machined components rejected by inspection from 10 batches. Show this information on a graph using the horizontal axis for the batch number.

batch number							
1	2	3	4	5	6	7	8
3	9	4	7	6	2	4	8
number of rejects							

9.6 A forging is taken from a furnace and left to cool in air. The table shows the measured temperature of the forging as it cools over a period of one hour. Draw a graph to show this information using the horizontal axis for time.

time in minutes						
0	10	20	30	40	50	60
900	610	390	265	195	140	100
temperature °C						

9.7 The table gives the circumference of circles over a range of diameters from 0 to 49 cm.

diameter cm							
0	7	14	21	28	35	42	49
0	22						
circumference cm							

a) Complete the table.
b) Show the information on a graph, use the x-axis for circumference.
c) Estimate the circumference of the following circles:
   (i) 30 cm diameter   (ii) 45 cm diameter

9.8 The table shows the results of an experiment to find the force required to overcome friction when sliding blocks of different masses along a horizontal surface.

mass of block in kg						
4	5	6	7	8	9	10
7.8	9.8	11.7	13.7	15.6	17.5	20
force in newtons						

a) Show these results on a graph using the horizontal axis for mass.
b) Estimate the force required to slide a block of mass 6.5 kg.

9.9 In an electrical experiment the current in amps and the voltage were recorded as shown in the table.

current in amps					
0.2	0.4	0.6	0.8	1.0	1.2
12	24	35	49	60	73
voltage in volts					

a) Plot the graph with voltage on the vertical axis.
b) Estimate the voltage for a current of 0.5 amps.

9.10 Draw conversion graphs for each of the following. The first-named quantity in each case is to be on the horizontal axis:
a) Inches and millimetres, for the range 0 to 8 inches.
b) Diameter and area of circles, for the range 0 to 10 cm diameter.

c) Degrees Fahrenheit and degrees Celsius, for the range 40°F to 100°F.
[$C = \frac{5}{9}(F - 32)$]

9.11 The table shows the results of a machining test to measure the life of a cutting tool before breakdown over a range of cutting speeds.

cutting speed in m/min				
26	28	30	35	38
60	40	23	8	4
tool life in minutes				

a) Plot the values on a graph using the horizontal axis for cutting speed, and join the points with a smooth curve.
b) From the graph estimate the predicted life of the cutting tool at a cutting speed of 32 m/min.

9.12 The table shows the spindle speed of a lathe in rev/min when machining a range of work diameters at a cutting speed of 66 m/min.

work diameter in mm						
20	30	40	50	60	70	80
1050						
spindle speed in rev/min						

a) Complete the table using the formula

$$N = \frac{1\,000\,V}{\pi d}$$

b) Plot the graph using the horizontal axis for work diameter. Join the points with a smooth curve.
c) Find the spindle speed required to machine a diameter of 33 mm.

9.13 Draw the graphs of the following equations for values of x from 0 to 6.

a) $y = 1.5x$
b) $y = 2x + 1$
c) $y = x^2$
d) $y = 3x - 2$
e) $y = x^2 + x$
f) $y = 0.5x + 0.7$
g) $y = \frac{x}{2} + 0.4$
h) $y = \frac{3x}{2} - 1.2$

9.14 Draw conversion graphs for each of the following, with the first-named quantity in each case to be on the horizontal axis.

a) gallons and litres, for the range 0 to 10 gallons
b) kilogrammes and pounds, for the range 0 to 80 kilogrammes

c) mass in kilogrammes and weight in newtons, for the range 0 to 1 000 kilogrammes

Use the conversion factors: 1 gallon = 4.55 litres
   1 kilogramme = 2.2 pounds
   1 kilogramme = 9.81 newtons

9.15   The table shows the voltage V volts across a lamp when the current flowing is I milliamperes

Voltage V volts									
0	1	2	3	4	5	6	7	8	9
0	139	207	260	317	360	396	434	466	498
Current I milliamperes									

a) Plot this information on a graph using the horizontal axis for voltage.
b) Find i) the current when the voltage is 7.4 volts, ii) the voltage when the current is 340 milliamperes.

9.16   The following results were obtained for a linear resistor in an electrical circuit.

Voltage V volts										
0	1	2	3	4	5	6	7	8	9	10
0	43	86	129	172	215	258	301	344	387	430
Current I milliamperes										

a) Plot this information on a graph using the horizontal axis for voltage.
b) Find i) the voltage when the current is 360 milliamperes, ii) the current when the voltage is 2.8 volts.

9.17   The table shows the electrical power P in watts taken by an electrical circuit for various values of current I in amperes. Show this data on a graph using the horizontal axis for current.

I (amps)								
0	1	2	3	4	5	6	7	8
0	1.5	6	13.5	24	37.5	54	73.5	96
P (watts)								

9.18   The table shows the resistance in ohms of an electrical conductor over a range of temperatures in °C. Draw a graph of this information using the horizontal axis for temperature.

Temperature °C							
10	20	30	40	50	60	70	80
4.70	4.94	5.18	5.52	5.71	5.93	6.30	6.62
Resistance ohms							

9.19   A test on a non-linear resistor gave the following values.

V volts				
0	5	10	15	20
0	15	51	216	486
I milliamperes				

a) Show this information on a graph using the horizontal axis for voltage.
b) Estimate i) the current when the voltage is 16.5 volts, ii) the voltage when the current is 100 milliamperes.

9.20   The voltage drop across a resistor for various values of current flowing through the resistor is shown below.

Voltage (volts)					
8	16	24	32	40	48
0.2	0.4	0.6	0.8	1.0	1.2
Current (amps)					

Show this information on a graph using the horizontal axis for current.

9.21   The time taken to manufacture a component is made up as follows:

Presswork	30 min
Machining	70 min
Mechanical fitting	85 min
Electrical fitting	40 min
Testing	15 min

Show this information on a Pie Chart.

9.22   The total skilled labour force of a factory is 300 workers made up as follows:

Machinists	140
Mechanical fitters	50
Electrical fitters	40
Welders	30
Sheet metal workers	25
Inspectors	15

Show this information on a Pie Chart.

9.23   The table shows the percentage of accidents in a factory attributed to various causes.

Electrical	12%
Moving machinery	16%
Hand tools	24%
Falling objects	8%
Burns	15%
Other causes	25%

a) Show this information on a Pie Chart.
b) If the total number of accidents was 150 how many accidents occurred when using hand tools?

9.24   The Pie Chart shown in fig. 9.23 shows the power ratings of electric motors in a machine shop. If the total number of electric motors is 120, find

a) the number of motors below 1 kW rating
b) the number of motors above 2 kW rating
c) the percentage of motors above 10 kW rating.

Fig. 9.23

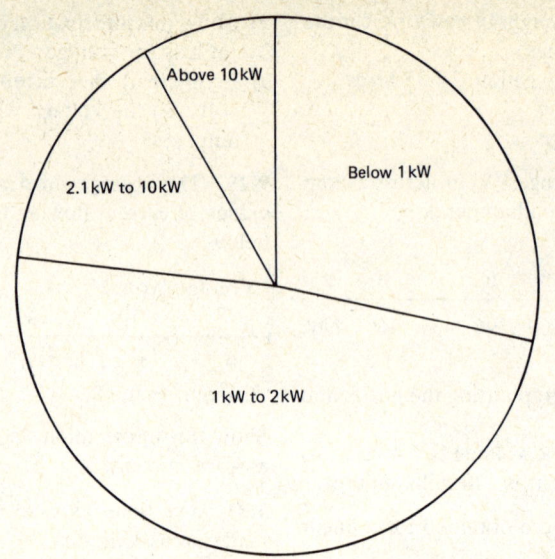

9.25 A certain component can be produced by 3 alternative machining methods. The time taken by each method is shown below.

Shaping   48 min
Milling   26 min
Planing   55 min

Show this information on a Bar Chart.

9.26 A special-purpose machine tool is equipped with 5 electric motors. The power ratings of the motors are shown below.

Main spindle motor   2.5 kW
Traverse motor   0.8 kW
Pump motor   0.4 kW
Milling head motor   1.5 kW
Milling feed motor   1.2 kW

Show this information on a Bar Chart.

9.27 The average number of breakdowns per machine over a 2 year period is shown below for various types of machine tools in a factory.

Types of machine tool	Average breakdowns
Lathes	4
Milling machines	7
Drilling machines	14
Shapers	9
Power saws	10
Grinding machines	8
Boring machines	2
Planing machines	1

Show this information on a Bar Chart.

9.28 The table shows the results of a series of tool life tests carried out over a range of cutting speeds.

Cutting speed, m/min	Tool life, min
16–18	10.5
18–20	16.5
20–22	21.2
22–24	29.4
24–26	32.5
26–28	28.7
28–30	26.4
30–32	20.5
32–34	17.6
34–36	9.2

Show this information on a Histogram.

9.29 A large manufacturer has 5 factories in different locations. The table shows the number of employees at each location.

Factory location	Number of employees
London	950
North West	650
Wales	250
Midlands	1 400
Scotland	350

Present this information in the form of *a*) a Pie Chart, *b*) a Bar Chart.

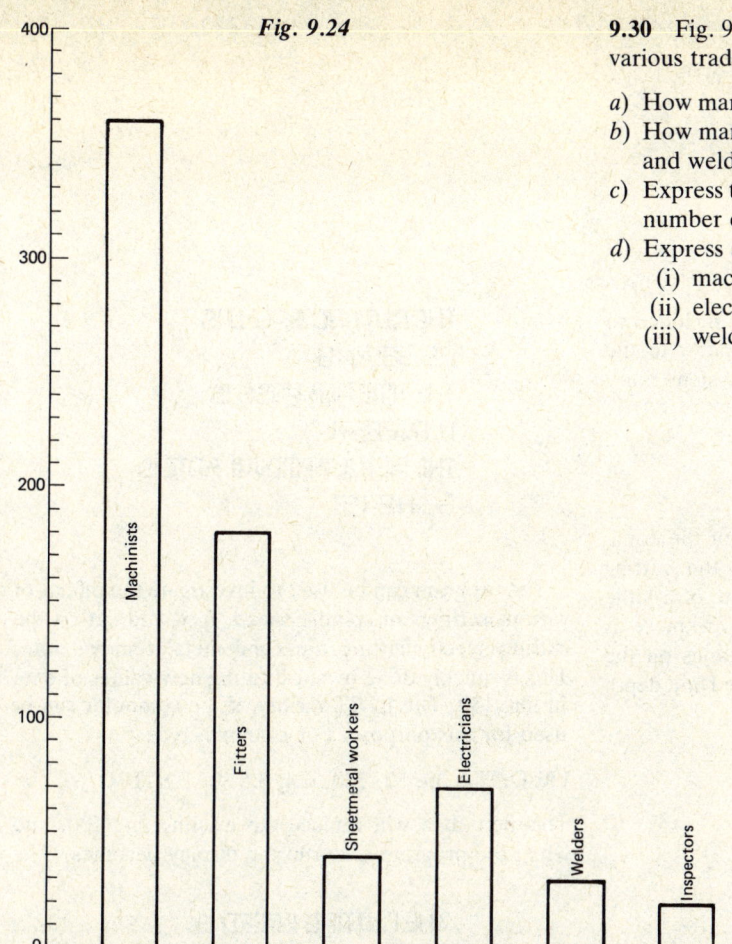

Fig. 9.24

9.30 Fig. 9.24 shows the number of skilled workers in various trades employed at a factory.

a) How many more fitters are employed than welders?
b) How many workers are employed in the sheetmetal and welding trades?
c) Express the number of inspectors as a fraction of the number of machinists.
d) Express as a percentage of the total work force:
   (i) machinists
   (ii) electricians
   (iii) welders

Appendix

The following ZX Spectrum programs are included to give examples of the use of a microcomputer in the solution of workshop problems. (Some elementary programs are shown in Chapter 4.)

PROGRAM 1 This program makes use of the equations derived in Chapter 8 to compute the cutting speed, cutting time and metal removal rate of a lathe operation in which the diameter of a workpiece is reduced in a single pass. The known settings on the machine are spindle speed, longitudinal feed and depth of cut.

Diameter of workpiece $D = 30$ mm
Length of pass $L = 80$ mm
Spindle speed $N = 280$ rev/min
Longitudinal feed $F = 0.4$ mm/rev
Depth of cut $C = 5$ mm

The values to be computed are
Cutting speed V m/min
Time per pass T min
Metal removal rate M mm^3/min

```
1Ø   REM "LATHE WORK"
2Ø   READ D
3Ø   READ L
4Ø   READ N
5Ø   READ F
6Ø   READ C
7Ø   LET V = PI * D * N/1ØØØ
8Ø   LET T = L/(F * N)
9Ø   LET M = PI * D * N * F * C
1ØØ  PRINT "THE CUTTING SPEED IS", V
11Ø  PRINT "THE TIME PER PASS IS", T
12Ø  PRINT "THE METAL REMOVAL RATE IS",
     M
13Ø  DATA 3Ø. 8Ø, 28Ø, Ø.4, 5
14Ø  STOP
```

Press RUN; press ENTER.

THE CUTTING SPEED IS
26.389378
THE TIME PER PASS IS
0.71428571
THE METAL REMOVAL RATE IS
52778.757

The program can be used to investigate the effects of various settings of spindle speed, feed and cut on the cutting speed, cutting time and metal removal rate. This is quickly done by substituting new values of data in line 13Ø. The EDIT facility of the computer can be used for this purpose. For example, type

13Ø DATA 3Ø, 8Ø, 32Ø, Ø.4, 3.5   ENTER

This new data will replace the existing line 13Ø, and when the program is re-run the display becomes:

THE CUTTING SPEED IS
30.159289
THE TIME PER PASS IS
T=0.625
THE METAL REMOVAL RATE IS
42223.005

Programs of this nature offer a rapid method of illustrating how standard measures of cutting efficiency are affected by changes in cutting variables.

PROGRAM 2 This program makes use of electrical formulae to compute the equivalent resistance, current flowing and power consumed in a circuit containing two resistors connected in parallel. The known quantities of the circuit are the supply voltage and the values of the two resistors.

Supply voltage $V = 12$ volts
Resistor $R1 = 10$ ohms
Resistor $R2 = 40$ ohms

The values to be computed are
Equivalent resistance Rp ohms
Current flowing in the circuit I amps
Power consumed in the circuit P watts

The electrical formulae to be used are

$$\frac{1}{Rp} = \frac{1}{R1} + \frac{1}{R2}$$

Hence $\quad Rp = \dfrac{R1 \times R2}{R1 + R2}$

Ohm's Law $\quad E = IR$
$P = EI$

```
1Ø   REM "PARALLEL CIRCUIT"
2Ø   READ R1
3Ø   READ R2
4Ø   READ E
5Ø   LET Rp = R1 * R2/(R1 + R2)
6Ø   LET I = E/Rp
7Ø   LET P = E * I
8Ø   PRINT "THE EQUIVALENT
     RESISTANCE IS", Rp
9Ø   PRINT "THE CURRENT FLOWING IS", I
1ØØ  PRINT "THE POWER CONSUMED IS", P
11Ø  DATA 1Ø, 4Ø, 12
12Ø  STOP
```

Press RUN, press ENTER.

```
THE EQUIVALENT RESISTANCE IS
8
THE CURRENT FLOWING IS
1.5
THE POWER CONSUMED IS
18
```

The program can be used to show the effect of varying the values of the resistors and the supply voltage by substitution of new data in line 11Ø. For example, type

11Ø   DATA 45Ø, 65Ø, 11Ø   ENTER

This new data will replace the existing line 11Ø, and when the program is re-run the display becomes

```
THE EQUIVALENT RESISTANCE IS
265.90909
THE CURRENT FLOWING IS
0.41367521
THE POWER CONSUMED IS
45.504274
```

PROGRAM 3 This program draws a bar chart from listed data. The values of x and y give the plotted position of the bottom left-hand corner of each bar. The value of b represents the width of each bar and is constant in this example. The value of H represents the height of each bar.

The table shows the number of faulty components produced in each of 10 batches of small turned parts. This information is to be shown on a bar chart.

Batch number									
1	2	3	4	5	6	7	8	9	10
40	65	75	30	55	20	105	75	65	25

Number of faulty parts

```
1Ø   REM "BAR CHART"
2Ø   READ x
3Ø   READ y
4Ø   READ b
5Ø   READ H
6Ø   IF x = 999 THEN STOP
7Ø   PLOT x, y
8Ø   DRAW b, Ø
9Ø   DRAW Ø, H
1ØØ  DRAW −b, Ø
11Ø  DRAW Ø, −H
12Ø  GO TO 2Ø
13Ø  DATA Ø, Ø, 15, 4Ø
14Ø  DATA 15, Ø, 15, 65
15Ø  DATA 3Ø, Ø, 15, 75
16Ø  DATA 45, Ø, 15, 3Ø
17Ø  DATA 6Ø, Ø, 15, 55
18Ø  DATA  75, Ø, 15, 2Ø
19Ø  DATA 9Ø, Ø, 15, 1Ø5
2ØØ  DATA 1Ø5, Ø, 15, 75
21Ø  DATA 12Ø, Ø, 15, 65
22Ø  DATA 135, Ø, 15, 25
23Ø  DATA 999, 999, 999, 999
```

Press RUN; press ENTER.

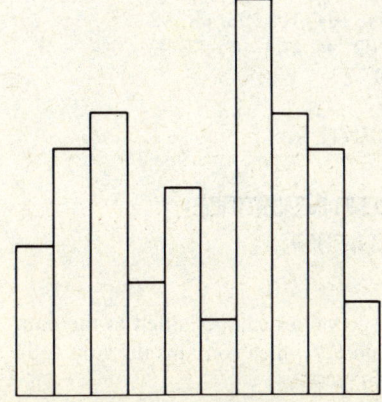

In line 6∅, the value x = 999 is chosen as an impossible value so that it may be used later in line 23∅ as a device for stopping the program.

This program may be used to draw bar charts of other information by substituting new values of H in the data lines. The program may also be used to draw histograms by substitution of suitable values of x, b and H.

PROGRAM 4 This program computes the mean or average value of a set of listed data. C represents the count of the values R, and S is the sum of the values. Because S = 0 when C = 1, the number of values n = C − 1, and the mean value is given by:

$$\frac{\text{Sum of the values}}{\text{Number of values}} = \frac{S}{n}$$

Example Find the mean resistance of a batch of resistors having the following measured values of resistance.

103, 98, 99, 100, 95, 96, 93, 94, 96, 100, 102, 93, 97, 98, 91, 96, 97, 99, 98, 104, 106, 102, 97, 93, 95, 101, 99, 95, 91, 94, 104, 105, 98, 98, 92, 90, 102, 94, 100, 97, 97, 93

```
1∅    REM "MEAN RESISTANCE"
2∅    LET C = 1
3∅    LET S = ∅
4∅    READ R
5∅    IF R = −999 THEN GO TO 9∅
6∅    LET S = S + R
7∅    LET C = C + 1
8∅    GO TO 4∅
9∅    LET n = C − 1
1∅∅   PRINT "MEAN RESISTANCE =", S/n
11∅   DATA 1∅3, 98, 99, 1∅∅, 95, 96, 93
12∅   DATA 94, 96, 1∅∅, 1∅2, 93, 97, 98
13∅   DATA 91, 96, 97, 99, 98, 1∅4, 1∅6
14∅   DATA 1∅2, 97, 93, 95, 1∅1, 99, 95
15∅   DATA 91, 94, 1∅4, 1∅5, 98, 98, 92
16∅   DATA 9∅, 1∅2, 94, 1∅∅, 97, 97, 93
17∅   DATA −999
```

Press RUN; press ENTER.

**MEAN RESISTANCE =
97.428571**

Any further measured values can be added to the data but must precede line 17∅ which contains the stop code −999. For example,

161   DATA 99, 1∅6, 98, 1∅5, 99, 97, 96

When the program is re-run the display becomes:

**MEAN RESISTANCE =
97.795918**

This is a useful workshop program because it may be used at any intervals during a production run to compute the mean level at which the product is being produced. The value of the mean can be constantly updated with data of the latest measured values, or earlier values can be deleted from the data.

Answers

Exercises 1

1.1   *a*) 67.21   *b*) 15.322   *c*) 18.0259   *d*) 157.45   *e*) 3.4467

1.2   *a*) 51.69   *b*) 147.36   *c*) 7.732   *d*) 25.487   *e*) 79.3335

1.3   *a*) 4.466   *b*) 38.5352   *c*) 211.65   *d*) 3.51   *e*) 8048.7

1.4   *a*) 89   *b*) 3.1   *c*) 24.5   *d*) 65.83   *e*) 6.09

1.5   11.07

1.6   *a*) 8.433   *b*) 8.43

1.7   *a*) 1.893   *b*) 0.04   *c*) 8.0652   *d*) 79.68   *e*) 0.0354

1.8   130.5 mm   **1.9**   108.2 mm   **1.10**   87.57 mm

1.11 27.32 mm   **1.12** $A = 85.27$ mm, $B = 78.64$ mm

1.13 76.3 mm   **1.14**   82.8 mm   **1.15**   36.8 mm

1.16 $a = 41.9$ mm, $b = 33.9$ mm, $c = 28.4$ mm, $d = 35.8$ mm

1.17 90.32 mm   **1.18**   17.05 mm

1.19 $a = 76.2$ mm, $b = 40.64$ mm, $c = 325.12$ mm

1.20 141 375 mm³   **1.21**   2621.1 mm²

1.22 *a*) 42.8 mm   *b*) 11

1.23 0.3 mm   **1.24**   *a*) 3.5 mm   *b*) 19   **1.25**   74

1.26 11.97 kW   **1.27**   6.3 kW

1.28 *a*) 27 ohms   *b*) 5.5 ohms

1.29 *a*) 140 V   *b*) 6.75 V   *c*) 18.75 V

1.30 8.8, 5.25, 7.5, 10, 48

1.31 10, 50, 31.5, 4.5, 21

1.32 33, 43.33, 13.63, 12.27, 11.11

1.33 *a*)   i) 0.2   ii) 0.75   iii) 1.35   iv) 0.165   v) 0.075   *b*)   i) 500   ii) 1340   iii) 285   iv) 80   v) 1057

1.34 *a*)   i) 0.65   ii) 0.075   iii) 2.38   iv) 0.175   v) 0.4   *b*)   i) 100   ii) 250   iii) 870   iv) 1300   v) 10 500

1.35 *a*) 2.1   *b*) 3.85   *c*) 2.75   *d*) 0.67   *e*) 4.07   *f*) 3.975   *g*) 1.085   *h*) 0.125   *i*) 5.2   *j*) 1.75

Exercises 2

2.1   *a*) $\frac{6}{7}$ *b*) $\frac{10}{13}$ *c*) $\frac{19}{21}$ *d*) $\frac{15}{16}$ *e*) $1\frac{3}{5}$ *f*) $2\frac{1}{4}$ *g*) $5\frac{5}{8}$ *h*) $8\frac{3}{16}$

2.2   *a*) $\frac{1}{2}$ *b*) $\frac{3}{4}$ *c*) $2\frac{8}{21}$ *d*) $2\frac{2}{3}$ *e*) $2\frac{3}{4}$

2.3   *a*) $\frac{10}{7}$ *b*) $\frac{35}{12}$ *c*) $\frac{23}{4}$ *d*) $\frac{39}{8}$ *e*) $\frac{171}{16}$

2.4   *a*) $3\frac{1}{7}$ *b*) $3\frac{3}{4}$ *c*) $2\frac{3}{4}$ *d*) $2\frac{1}{8}$ *e*) $14\frac{5}{9}$

2.5   *a*) $\frac{3}{4}$ *b*) $\frac{1}{5}$ *c*) $\frac{2}{5}$ *d*) $\frac{13}{15}$ *e*) $\frac{11}{4}$ *f*) $\frac{49}{60}$ *g*) $\frac{2}{3}$ *h*) $\frac{3}{11}$

2.6   *a*) $\frac{5}{8}$ *b*) $\frac{7}{8}$ *c*) $\frac{15}{16}$ *d*) $\frac{17}{20}$ *e*) $\frac{13}{21}$ *f*) $1\frac{2}{21}$ *g*) $\frac{3}{4}$ *h*) $1\frac{17}{20}$ *i*) $1\frac{7}{8}$ *j*) $1\frac{11}{24}$ *k*) $1\frac{5}{6}$ *l*) $1\frac{17}{18}$

2.7   *a*) $2\frac{3}{8}$ *b*) $3\frac{3}{4}$ *c*) $4\frac{3}{20}$ *d*) $5\frac{11}{16}$ *e*) $9\frac{1}{8}$ *f*) $7\frac{3}{4}$ *g*) $12\frac{3}{8}$ *h*) $18\frac{1}{20}$

2.8   *a*) $\frac{3}{8}$ *b*) $\frac{5}{6}$ *c*) $\frac{19}{20}$ *d*) $\frac{7}{8}$ *e*) $\frac{3}{4}$ *f*) $1\frac{5}{8}$ *g*) $1\frac{3}{4}$ *h*) $\frac{1}{2}$ *i*) $1\frac{5}{6}$ *j*) $3\frac{7}{9}$

2.9   *a*) $\frac{1}{2}$ *b*) $\frac{8}{35}$ *c*) $\frac{10}{27}$ *d*) $\frac{6}{25}$ *e*) $\frac{63}{104}$ *f*) $\frac{5}{42}$ *g*) $\frac{1}{30}$ *h*) $\frac{5}{28}$

2.10 *a*) $1\frac{3}{8}$ *b*) $1\frac{1}{12}$ *c*) $1\frac{4}{5}$ *d*) $4\frac{3}{4}$ *e*) $3\frac{5}{7}$ *f*) $1\frac{1}{5}$ *g*) $1\frac{3}{4}$ *h*) $10\frac{1}{2}$

2.11 *a*) 2   *b*) $\frac{1}{2}$ *c*) $1\frac{1}{20}$ *d*) $2\frac{1}{4}$ *e*) $1\frac{1}{6}$ *f*) $1\frac{5}{22}$ *g*) $\frac{6}{17}$ *h*) $4\frac{7}{12}$

2.12 *a*) 4   *b*) $4\frac{1}{2}$ *c*) 9   *d*) $9\frac{3}{4}$ *e*) $1\frac{5}{6}$ *f*) $1\frac{13}{20}$ *g*) $3\frac{3}{5}$ *h*) $\frac{3}{4}$

2.13 170

2.14 *a*) 20   *b*) 10   *c*) 70

2.15 *a*) $\frac{1}{10}$ *b*) 50

2.16 *a*) 40 m   *b*) 120 kg   *c*) 60 cm   *d*) 224 g   *e*) 36 mm

2.17 $21\frac{1}{8}$ m²

2.18 length = 57 cm, wingspan = 66 cm

2.19 $\frac{11}{18}$

2.20 *a*) $\frac{47}{50}$ *b*) 61

2.21 $3\frac{3}{32}$ m³

2.22 *a*) 0.75   *b*) 1.8   *c*) 0.625   *d*) 2.375   *e*) 0.5625   *f*) 0.85   *g*) 8.26   *h*) 0.675

2.23 *a*) 0.538   *b*) 1.667   *c*) 3.833   *d*) 5.286

2.24 *a*) $\frac{7}{10}$ *b*) $\frac{22}{25}$ *c*) $\frac{11}{250}$ *d*) $\frac{7}{40}$ *e*) $2\frac{12}{25}$ *f*) $5\frac{3}{200}$ *g*) $7\frac{91}{250}$ *h*) $4\frac{21}{200}$

2.25 *a*) 0.922   *b*) 0.175   *c*) 1.954   *d*) 2.1951   *e*) 2.05   *f*) 3   *g*) 1.2495   *h*) 1.225

2.26 *a*) 1500   *b*) 2750   *c*) 4250   *d*) 2800   *e*) 3667

2.27 $14\frac{3}{4}$ kW

2.28 *a*) 1600   *b*) 5333   *c*) 750   *d*) 9200   *e*) 2250

2.29 $7\frac{1}{2}$ ohms

2.30 2, $1\frac{1}{3}$, 40, 8, $1\frac{7}{8}$

2.31 $7\frac{5}{12}\,\Omega$

2.32 $6\,\Omega$

Exercises 3

3.1 *a*) 60% *b*) 85% *c*) 36% *d*) 25%
e) 87.5% *f*) 68.75%

3.2 *a*) 40% *b*) 83% *c*) 17% *d*) 35.5%
e) 70.8% *f*) 1.5%

3.3 *a*) $\frac{3}{25}$ *b*) $\frac{16}{25}$ *c*) $\frac{7}{20}$ *d*) $\frac{3}{8}$ *e*) $\frac{81}{100}$ *f*) $\frac{37}{40}$

3.4 *a*) 0.16 *b*) 0.74 *c*) 0.235 *d*) 0.415
e) 0.07 *f*) 0.8025

3.5 *a*) 150 mm *b*) 60 kg *c*) 72 l *d*) 30 kg
e) 261 cm *f*) 7 m *g*) 42.6 kg *h*) 180 g

3.6 *a*) 5% *b*) 40% *c*) 9.5% *d*) 68%

3.7 *a*) $\frac{22}{25}$ *b*) 48 kg

3.8 18

3.9 *a*) 3% *b*) 5.4% *c*) 91.6%

3.10 *a*) 70% *b*) 30%

3.11 *a*) 14 *b*) 42 *c*) 50.16 *d*) 0.41

3.12 16.22 **3.13** 190 **3.14** 50.4

3.15 455 **316** 27 min **3.17** 2 h 52 min

3.18 12.02 mm

3.19 *a*) 1:4 *b*) 1:3 *c*) 2:13 *d*) 3:8 *e*) 1:4
f) 1:2:5 *g*) 1:2:4 *h*) 4:7:13

3.20 *a*) 1:9 *b*) 7:8 *c*) 11:19 *d*) 2:5:7

3.21 5:9

3.22 *a*) 7:9 *b*) 54 min

3.23 32 kg

3.24 *a*) 1.5 l *b*) 4 l

3.25 *a*) 12 kg *b*) 35 kg *c*) 7.5 kg

2.26 $A = 9.6$ kg, $B = 14.4$ kg, $C = 24$ kg

3.27 *a*) 2:5:7 *b*) 80 kg, 200 kg, 280 kg

3.28 1 h 45 min, 1 h, 15 min

3.29 *a*) 4 m *b*) 37.5 m

3.30 $B = 660$ rev/min, $C = 990$ rev/min

3.31 275 Ω, 225 Ω

3.32 81.6 V, 78.4 V

3.33 16, 20, $37\frac{1}{2}$, 20, $31\frac{1}{4}$

3.34 10 Ω, 25 Ω, 35 Ω

3.35 92

Exercises 4

4.1 *a*) 1170 *b*) 165 *c*) 831

4.2 *a*) 45 *b*) 5 *c*) 757

4.3 *a*) 15.21 *b*) 3.182 *c*) 69.82 *d*) 1069
e) 5367 *f*) 161400 *g*) 0.5184 *h*) 0.001962
i) 0.4202 *j*) 0.0009980

4.4 *a*) 1.16 *b*) 1.00 *c*) 28.16 *d*) 0.24

4.5 24.24

4.6 *a*) 6.928 *b*) 1.612 *c*) 10.95 *d*) 0.7874
e) 2.890 *f*) 7.192 *g*) 15.77 *h*) 0.5655
i) 86.21 *j*) 0.1378

4.7 *a*) 1.39 *b*) 0.94 *c*) 3.53 *d*) 0.28

4.8 *a*) 7.672 *b*) 6.982 *c*) 1.351 *d*) 343.592

4.9 *a*) 7.92 cm *b*) 27.78 cm *c*) 4.770 cm
d) 61.45 mm *e*) 15.82 cm *f*) 435.3 mm

4.10 131.1 mm **4.11** 791.3 mm **4.12** 814 mm

4.13 41.00 mm **4.14** 28.28 mm **4.15** 335.9 mm

4.16 135.4 mm **4.17** 226 mm **4.18** 4.610

4.19 5.552 **4.20** 5984 **4.21** 25.17

4.22 0.6381 **4.23** 0.03966 **4.24** 2641

4.25 1.495 **4.26** 277.5 **4.27** 0.01033

4.28 2.226 **4.29** 0.8076 **4.30** 13.43

4.31 199.6 **4.32** 1.491 **4.33** 8.041

4.34 396.3 **4.35** 89.45 **4.36** 12.51

4.37 47.80 **4.38** 4007 **4.39** 4.942

4.40 0.5301 **4.41** 403.0117 **4.42** 21.93044

4.43 8317.0344 **4.44** 1555.7582 **4.45** 0.067211

4.46 2.232 **4.47** 5.08645 **4.48** 46782.792

4.49 0 **4.50** 85.4

4.51 549.8, 125.7, 443.0, 188.5, 1093

4.52 96390, 27870, 59060, 440800, 150800

4.53 29.1, 34.6, 43.4, 30.7, 33.9

4.54 17.32, 20.78, 25.98, 43.30, 65.82

4.55 24.86 N **4.56** 2124 W

4.57 i) 4.472 A ii) 8.944 A iii) 5.123 A
iv) 9.487 A v) 7.303 A

4.58 0.4, 0.14, 0.57, 0.34, 1.18, 0.63

4.59 10.02 mm² **4.60** 13.59 ohms

4.61 *a*) 95.15 *b*) 5320000 *c*) 41600 *d*) 5.588
e) 38.86 *f*) 2.122 *g*) 0.3344 *h*) 891000
i) 1.928 *j*) 41.38 *k*) 297.9 *l*) 78.41
m) 981.9 *n*) 2.610

4.62 $I = 40.0$, 0.214, 4.76, 1.50, 4.17
$P = 800$, 1.61, 238, 18.0, 417

Exercises 5

5.1 *a*) 3 780 m   *b*) 1 043 mm   *c*) 8.9 cm
   d) 2.64 m   *e*) 130 cm   *f*) 4.965 km
   g) 950 mm   *h*) 6.75 m

5.2 *a*) 93.98 mm   *b*) 10.9728 m   *c*) 23.114 mm
   d) 163.83 mm   *e*) 24.1395 km   *f*) 88.9 mm
   g) 36.83 cm   *h*) 0.4572 m

5.3 *a*) 1.5 inches   *b*) 3.6 inches   *c*) 0.55 inch
   d) 12.5 inches   *e*) 4.95 inches   *f*) 2.5 inches
   g) 14 inches   *h*) 0.3 inch

5.4 *a*) 6.09 mm   *b*) 219.20 mm   *c*) 44.58 mm
   d) 120.65 mm   *e*) 20.75 mm   *f*) 78.13 mm
   g) 0.13 mm   *h*) 397.71 mm

5.5 lengths: 165.1 mm, 38.1 mm, 31.8 mm
   diameters: 69.9 mm, 57.2 mm, 19.1 mm

5.6 lengths: 95.25 mm, 36.78 mm, 25.78 mm
   heights: 70.03 mm, 33.50 mm, 20.32 mm

5.7 67.33 mm   **5.8** 95.20 mm

5.9 *a*) 35.34 mm   *b*) 22.20 mm   *c*) 52.51 mm
   d) 44.47 mm

5.10 9.3 mm^2

5.11 *a*) 1.4 km^2 *b*) 7.3 m^2 *c*) 2.918 m^2
   d) 4 500 000 m^2 *e*) 10 cm^2 *f*) 14 700 mm^2

5.12 *a*) 4 064.51 mm^2 *b*) 2.17 inches^2 *c*) 54.84 cm^2

5.13 *a*) 300 mm^2, 80 mm   *b*) 29.25 mm^2, 22 mm
   c) 61 200 mm^2, 1 040 mm   *d*) 1.4 m^2, 5.1 m
   e) 234.5 mm^2, 61.8 mm   *f*) 4 424 mm^2, 270 mm
   g) 127 795 mm^2, 1 448 mm   *h*) 2.73 mm^2, 7.3 mm

5.14 *a*) 23 625 mm^2 *b*) 62 500 mm^2 *c*) 400 mm^2
   d) 126 m^2 *c*) 95 cm^2 *f*) 3.2766 m^2 *g*) 1.2 m^2
   h) 3 575 mm^2

5.15 *a*) i) 40 m^2, 26 m   ii) 25 m^2, 20 m   iii) 20 m^2,
   18 m   iv) 48 m^2, 32 m   v) 156 m^2, 50 m
   b) 289 m^2

5.16 205 000 mm^2 **5.17** 2 035 mm^2

5.18 *a*) 72 mm^2 *b*) 384 mm^2 *c*) 99.84 mm^2
   d) 1 008 mm^2 *e*) 1 800 mm^2 *f*) 1 480 mm^2

5.19 1.96 mm^2

5.20 *a*) 56 mm^2 *b*) 127.5 mm^2 *c*) 675 mm^2
   d) 58.5 mm^2 *e*) 15.275 mm^2 *f*) 437 mm^2
   g) 126.04 mm^2 *h*) 14 798 mm^2

5.21 *a*) 126.5 mm^2 *b*) 1 760 mm^2 *c*) 31.5 cm^2
   d) 2 560 mm^2 *e*) 0.735 m^2 *f*) 43 472 mm^2
   g) 48.825 mm^2 *h*) 1.04 mm^2

5.22 *a*) 235 600 mm^2 *b*) 866 mm^2 *c*) 8 250 mm^2
   d) 3 200 mm^2

5.23 *a*) 264 mm   *b*) 616 mm   *c*) 462 mm
   d) 132 cm   *e*) 1 100 cm   *f*) 11 m   *g*) 33 mm
   h) 495 mm

5.24 *a*) 104.31 mm   *b*) 588.81 mm   *c*) 10.02 cm
   d) 45.87 mm   *e*) 79.81 mm   *f*) 102.43 mm
   g) 6.85 mm   *h*) 154.90 mm

5.25 *a*) 154 m^2 *b*) 346.5 mm^2 *c*) 1.54 mm^2
   d) 962.5 mm^2 *e*) 616 cm^2 *f*) 67 914 mm^2
   g) 385 000 mm^2 *h*) 18 634 mm^2

5.26 *a*) 1 164 mm^2 *b*) 169 100 mm^2 *c*) 12 230 mm^2
   d) 2 790 cm^2 *e*) 1.674 m^2 *f*) 7 015 mm^2
   g) 5 411 mm^2 *h*) 211 600 mm^2

5.27 659.8 mm^2 **5.28** 3 150 mm^2

5.29 2 392 mm^2 **5.30** 14 750 mm^2

5.31 0.1118, 0.1575, 0.2489, 0.3277, 0.5080

5.32 18 380 mm^2

Exercises 6

6.1 *a)* 15 600 mm³ *b)* 1 800 cm³ *c)* 31 500 mm³
*d)*15 552 mm³ *e)* 10 656 mm³ *f)* 33 750 mm³

6.2 *a)* 240 000 mm³ *b)* 8 670 mm³ *c)* 23 616 mm³
d) 3 430 mm³ *e)* 180 000 mm³
f) 312 500 mm³ *g)* 2 220 000 mm³ *h)* 222.3 mm³

6.3 *a)* 800 000 mm³ *b)* 1.47 m³ *c)* 3 600 cm³
d) 1 250 000 mm³ *e)* 288 000 mm³
f) 1 400 000 mm³ *g)* 24 000 000 mm³
h) 16 200 cm³

6.4 *a)* 4 813 mm³ *b)* 5 119 mm³ *c)* 6 431 mm³
d) 8 881 mm³

6.5 *a)* 9 000 mm² *b)* 16 500 mm² *c)* 337 500 mm²

6.6 *a)* 160 000 mm³ *b)* 140 800 mm³
c) 180 000 mm³ *d)* 81 600 mm³
e) 400 000 mm³ *f)* 616 000 mm³

6.7 3 850 000 mm³ **6.8** 975 000 mm³

6.9 *a)* 9 240 mm³ *b)* 385 cm³ *c)* 138 600 mm³
d) 3.08 m³ *e)* 962 500 mm³ *f)* 194 040 mm³

6.10 *a)* 5 400 mm³ *b)* 4 330 cm³ *c)* 2.12 m³
d) 135 000 mm³ *e)* 1 550 mm³ *f)* 5.03 mm³

6.11 *a)* 123 000 mm³ *b)* 57 800 mm³ *c)* 42 600 mm³

6.12 32 800 mm³

6.13 *a)* 200 000 mm³ *b)* 79 000 mm³ *c)* 694 000 mm³

6.14 *a)* 13 840 mm³ *b)* 65 000 mm³ *c)* 74 960 mm³

6.15 113 mm³ **6.16** 7 **6.17** 75 cm³

6.18 *a)* 8 000 mm³ *b)* 46 200 mm³ *c)* 1 437 cm³
d) 100 000 mm³ *e)* 6 284 mm³ *f)* 80 cm³

6.19 13 000 mm³ **6.20** 35 000 mm³

6.21 *a)* 3 m³ *b)* 2.7 m³ *c)* 1.54 m³ *d)* 0.816 m³
e) 2.1 m³ *f)* 5.156 m³ *g)* 0.0945 m³
h) 1.14 m³

6.22 3 959 cm³

6.23 *a)* 2121 *b)* 271 *c)* 11 *d)* 31 *e)* 120961
f) 211

6.24 78 550 mm³

6.25 *a)* 28 500 mm³ *b)* 33.9%

6.26 160 000 mm³ **6.27** 354 000 mm³

6.28 *a)* 56 600 mm³ *b)* 18 100 mm³ *c)* 39 200 mm³
d) 12 100 mm³

6.29 *a)* 7 200 mm² *b)* 4 032 mm² *c)* 6 900 mm²
d) 10 032 mm²

6.30 *a)* 10 560 mm² *b)* 1 257 mm² *c)* 1 100 cm²

6.31 *a)* 900 000 mm³ *b)* 210 600 mm²

6.32 *a)* 3 m³ *b)* 2.686 m³

6.33 346.41

6.34 46 650, 6 371, 22 580

6.35 *a)* 75.4 *b)* 0.147 *c)* 3.56

6.36 *a)* 74 220 *b)* 6 283 *c)* 1 257 000

6.37 *a)* 13 520 000 mm³ *b)* 28 800 000 mm³

6.38 31 260 mm³ **6.39** 150.51 **6.40** 4 860 mm³

Exercises 7

7.1 *a)* 234′ *b)* 435″ *c)* 4 200″ *d)* 762′
e) 7 754″ *f)* 16 500″ *g)* 150′ *h)* 345′

7.2 *a)* 1°50′ *b)* 3′20″ *c)* 1°6′40″ *d)* 1°30′30″
e) 5°50′ *f)* 11′40″ *g)* 1°47′30″ *h)* 4°26′40″

7.3 *a)* 14°30′ *b)* 31°27′ *c)* 18°39′ *d)* 7°16′48″
e) 13°25′12″ *f)* 118°05′24″ *g)* 28°09′36″
h) 130°43′48″

7.4 *a)* 2.5° *b)* 2.6° *c)* 18.7° *d)* 33.25°
e) 8.42° *f)* 19.64° *g)* 104.78° *h)* 20.225°

7.5 *a)* 91°31′ *b)* 145°34′ *c)* 1°37′22″
d) 23°13′18″ *e)* 45°40′27″ *f)* 101°10′18″
g) 17°34′ *h)* 40°27′06″

7.6 *a)* 10°04′ *b)* 41.9°

7.7 *a)* 64° *b)* 10°49′ *c)* 37° *d)* 12°30′

7.8 76°13′

7.9 *a)* 60° *b)* 120° *c)* 55°

7.10 *a)* 106° *b)* 36°12′ *c)* 27°46′ *d)* 32°50′

7.11 *a)* 15° *b)* 27°25′ *c)* 46°32′46″ *d)* 38°49′25″
e) 23°20′ *f)* 29°

7.12 *a)* 40° *b)* 50° *c)* 84° *d)* 64° *e)* 32°
f) 38° *g)* 78° *h)* 85°

7.13 *a)* wedge angle 60° *b)* rake angle 15°
c) clearance angle 65° *d)* wedge angle 70°
e) clearance angle 5°30′ *f)* rake angle 22°
g) rake angle 3°30′ *h)* wedge angle 57°

7.14 *a)* 18° *b)* 20° *c)* 17°

7.15 Acute angles: 48°, 85°, 60°
Obtuse angles: 130°, 175°, 100°
Reflex angles: 200°, 300°

7.16 *a)* acute *b)* reflex *c)* obtuse
d) supplementary *e)* complementary
f) supplementary *g)* reflex *h)* acute

7.17 *a)* 70°, 110°, 110° *b)* 125°, 55°, 125°
c) 117°, 63°, 63° *d)* 19°30′, 160°30′, 160°30′
e) 32°, 148°, 148° *f)* 82°37′, 97°23′, 97°23′

7.18 *a)* 45° *b)* 135° *c)* 45° *d)* 45° *e)* 135°
f) 45° *g)* 135°

7.19 *x* = 86°32′, *y* = 86°32′

7.20 *x* = 65°, *y* = 115°, *z* = 65°

7.21 *a)* 31° *b)* 39° *c)* 13° *d)* 26° *e)* 40°20′
f) 61°

7.22 *a)* 75° *b)* 67°30′ *c)* 50° *d)* 18° *e)* 40°
f) 61°

7.23 clearance angle = 20°, rake angle = 10°

7.24 *x* = 130°, *y* = 50°

7.25 *a)* *x* = 15°, *y* = 35° *b)* *x* = 60°, *y* = 30°
c) *x* = 40°, *y* = 130° *d)* *x* = 120°, *y* = 120°

7.26 *a)* 40° *b)* 33° *c)* 38° *d)* 21°40′

7.27 *x* = 90°, *y* = 30°

7.28 80 mm

7.29 *a*) 320.8 mm² *b*) 20.53 cm² *c*) 104.7 mm²
7.30 *a*) $x = 88°$ *b*) $x = 100°$, $y = 97°$
   c) $x = 47°30'$, $y = 132°30'$
   d) $x = 112°$, $y = 107°$, $z = 35°$
   e) $x = 104°$, $y = 66°$, $z = 100°$ *f*) $x = 63°$
7.31 *a*) 9° *b*) 243°
7.32 32 divisions **7.33** 72°
7.34 101°30' **7.35** 16°22'

Exercises 8

8.1 i) $a + b - c$ ii) ac iii) $a + b + c$

iv) abc v) $\dfrac{a + c}{b}$ vi) $2a + 3c$

vii) $3b - \dfrac{c}{2}$ viii) $\dfrac{3a}{4} + \dfrac{b + c}{2}$

ix) $\dfrac{a + b + c}{3}$ x) $a^2 + b^2 + c^2$

8.2 *a*) xy *b*) $\dfrac{xy}{2}$ *c*) xy

d) $2xy$ *e*) $6xy$ *f*) $2xy$ *g*) $3xy$ *h*) $5xy$
8.3 *b*) £39.50
8.4 *a*) $10x$ *b*) $9y$ *c*) $2a + 2b$ *d*) $4x + 6y$
   e) not possible *f*) $10x + 6x^2$ *g*) $2x + 2x^2$
   h) $8a + 2a^2 + 2a^3$
8.5 *a*) $6b^2$ *b*) $21a^3$ *c*) $20y^3$ *d*) $6x^4$ *e*) $40a^3$
   f) $4a^3$ *g*) $6ab$ *h*) $96xyz$
8.6 *a*) $15a$ *b*) $4y$ *c*) $9a$ *d*) $12m^2$ *e*) $3a$
   f) $4ac$ *g*) $5xy^2$ *h*) $5a^2x$
8.7 *a*) $3a + 3b$ *b*) $6x + 8$ *c*) $24a + 4b$
   d) $20x - 30y$ *e*) $28x^2 + 7y$ *f*) $6a^2 - 3b^2$
   g) $x + xy$ *h*) $20a^2 - 8b^2 + 4c$ *i*) $x^2 + xy$
   j) $2xy - y^2$
8.8 *a*) $4x + 4$ *b*) $3a + 2$ *c*) $3a - 6$
8.9 i) 9 ii) 1 iii) 10 iv) 5 v) 14 vi) -2
   vii) 5 vii) -3
8.10 i) 8 ii) 22 iii) 40 iv) 18 v) 28 vi) -10
   vii) 2 viii) 5
8.11 *a*) 4 *b*) 6 *c*) 6 *d*) -16 *e*) -24 *f*) 4
   g) -1 *h*) -32
8.12 *a*) 13 *b*) 48 *c*) 27 *d*) 12 *e*) 46 *f*) 24
   g) 25 *h*) 283
8.13 *a*) 41.2 *b*) 100.48 *c*) 0.412 *d*) 0.12 *e*) 70
   f) 69 *g*) 0.0184 *h*) 0.18
8.14 *a*) 3 *b*) 14 *c*) 15 *d*) 5 *e*) 27 *f*) -2
   g) -3 *h*) 1.2 *i*) 3.5 *j*) 4.432 *k*) $1\frac{1}{4}$ *l*) $1\frac{7}{8}$
8.15 *a*) 10 *b*) 11 *c*) 6 *d*) 40 *e*) 125 *f*) 1
   g) $8\frac{1}{4}$ *h*) 4.582 *i*) 3 *j*) -4 *k*) $2\frac{1}{4}$ *l*) 2.1
8.16 *a*) 3 *b*) 6 *c*) 8 *d*) 3 *e*) 32 *f*) 3 *g*) 5
   h) 7
8.17 *a*) 10 *b*) 21 *c*) 2 *d*) 2 *e*) 6 *f*) 14.49
   g) 12.84 *h*) 0.001
8.18 *a*) 3 *b*) 4 *c*) 2 *d*) -1 *e*) $\frac{1}{2}$ *f*) 10 *g*) 12
   h) $1\frac{1}{4}$
8.19 *a*) 4 *b*) 8 *c*) 9 *d*) 0.3
8.20 *a*) 4 *b*) 8 *c*) $2\frac{1}{2}$ *d*) 5 *e*) 2 *f*) 4 *g*) 7
   h) 6
8.21 *a*) 10 *b*) -1 *c*) 6 *d*) 2 *e*) 2.7 *f*) 0.3

8.22 a) $t = \dfrac{F}{xy}$ b) $m = \dfrac{g}{h}$

   c) $r = \dfrac{s}{\pi h}$ d) $T = \dfrac{I}{PR}$

   e) $v = \dfrac{A}{\pi r^2}$ f) $l = \dfrac{4V}{\pi D^2}$

   g) $m = \dfrac{n+q}{P}$ h) $R = \dfrac{EI}{M}$

8.23 a) $x = \dfrac{y - ab}{a}$ b) $k = \dfrac{C + 100a}{100}$

   c) $n = \dfrac{M}{P} + 2$ d) $n = \dfrac{t - r + s}{s}$

   e) $x = G(p + q)$ f) $b = \dfrac{0.9 - xd}{x}$

8.24 4 ohms

8.25 a) 1 652°F   b) $\frac{5}{9}(F - 32)$

8.26 6.25 amps

8.27 a) $A = \frac{1}{2}bh$ b) $V = \pi r^2 l$

8.28 a) 180° − plan angle − trail angle   b) 133°

8.29 a) $L = a + b + \frac{1}{2}\pi r$ b) 200 mm

8.30 a) 276 rev/min   b) 1 273 rev/min
   c) 102 rev/min   d) 159 rev/min

8.31 a) $V = \dfrac{\pi dN}{1\,000}$ b) 1 414 m/min

8.32 a) 28.28 m/min   b) 27.33 m/min
   c) 24.5 m/min

8.33 23.57, 37.70, 32.99 m/mm

8.34 3.33 min

8.35 a) $T = \dfrac{L}{fN}$ b) 22.5 s

8.36 a) 1 591   b) 6 s

8.37 2 min 24 s

8.38 a) i) $V + IR$ ii) $\dfrac{E - V}{R}$ iii) $\dfrac{E - V}{I}$

   b) i) 2.3   ii) 1.5   iii) 0.8   iv) 9.075

8.39 50, 96, 1.85, 105, 40, 3.75

8.40 a) i) $\dfrac{P}{I^2}$ ii) $\sqrt{\dfrac{P}{R}}$

   b) i) 120   ii) 11.95   iii) 15

8.42 a) 1.08 kW   b) 100 W   c) 120 mA

8.43 a) 0.313 A   b) 767 Ω

8.44 a) 13.33 Ω   b) 43.2 W   c) 33.33 V

8.45 a) 0.053   b) 2.08   c) 100

Exercises 9

9.1 a) 6, 10, 2, 4, 8, 9

9.2 a) 16, 9, 4, 1, 0, 1, 4, 9
   b) i) 6.25, ii) 0.84

9.7 a) 44, 66, 88, 110, 132, 154
   c) i) 94 cm, ii) 141 cm

9.8 12.8 N   **9.9** b) 30 V   **9.11** b) 15 min

9.12 a) 700, 525, 420, 350, 300, 263
   c) 636 rev/min

9.15 b) i) 445 mA   ii) 4.6 V

9.16 b) i) 8.4 V   ii) 120 mA

9.19 b) i) 285 mA   ii) 11.8 V

9.23 b) 36

9.24 a) 34   b) 28   c) 8.3%

9.30 a) 150   b) 70   c) $\frac{1}{18}$ d) i) 51.4%   ii) 10%
   iii) 4.3%

Index